a History of Artillery

HART

a History of Artillery

Ian V Hogg

Special Artwork by
John Batchelor

HAMLYN
London · New York · Sydney · Toronto

Published by
The Hamlyn Publishing Group Limited
Astronaut House, Feltham, Middlesex, England
© The Hamlyn Publishing Group Limited, 1974

ISBN 0 600 31314 X

Printed in England by
Jarrold and Sons Limited, Norwich

Contents

1 The Black Art 9

2 From Art to Science 25

3 From Scientist to Engineer 45

4 The Age of Invention 77

5 Ready for the Fray 99

6 Battle is Joined 123

7 Intermission 149

8 Battle is Resumed 179

9 Decline but not Fall 217

Index 237

Acknowledgments 240

Preface

Some years ago, Sir Robert Watson Watt wrote an account of the development of radar which he was careful to call *One Story of Radar*; in his introduction he pointed out that he had selected that particular title since it would be quite possible to relate the development to a totally different story, due to the fact that the subject was so vast and complicated. It is unfortunate that Sir Robert beat me to it, since 'One Story of Artillery' would have been most pertinent here. As it is, I have called this book *A History of Artillery*, since it represents one point of view; a different writer would doubtless produce a totally different narrative within the same framework. The complete history of artillery will never be written; the subject is so involved and enormous that it would take a lifetime of dedicated work, and at the end of it no publisher would give it houseroom because of its bulk. Moreover much of it would be detail of interest only to a minority of knowledgeable readers, albeit detail of importance.

Such being the case, when one is constrained by the economic facts of life to compress the story into manageable size, it becomes necessary to select and discard, so that the eventual result, while sticking to essential facts, reflects the writer's opinions and predilections. It thus becomes inevitable that some readers will consider that items of equal importance have been neglected or omitted. This, I am afraid, is unavoidable, and this preface is by way of explaining such lack. The first time I ever wrote on the subject of artillery was to produce an eight-thousand word résumé of the development of guns during the Second World War, and afterwards I was taken to task by a Chelsea Pensioner for failing to mention 'his' gun, the 4·5-inch howitzer. From which I concluded that you could please some of the people all of the time, all of the . . . I made up a very good little phrase there, but then I found somebody had beaten me to that too.

I.V.H.

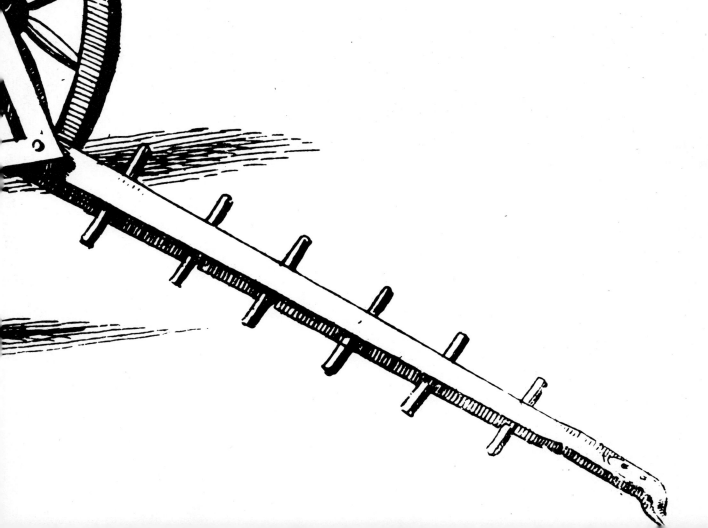

CHAPTER 1

The Black Art

*I*n spite of much research and writing over the past half-century the question of the invention of guns and gunpowder is still unresolved, and is likely to remain so. Assumptions can be made and approximations stated; but at least we can look closer at some of the claims which have been put forward and see how much validity they have.

It has been averred that the cannon was known in ancient times, even in the days of the Greeks and Romans, lost during the Dark Ages and re-discovered in the twelfth century. This theory seems to be based on a statement of Livy's, who described Archimedes as '*inventor ac machinator bellecorum*' and who reported that one of Archimedes' engines, 'with a terrible noise did shoot forth great bullets of stone'. Even greater antiquity is argued by the many people who have quoted the 'Gentoo Code' of laws, originating in India in about 1500 BC. In this text is a passage, under the heading 'Duty of the Magistrate', which was originally interpreted to read: 'He shall not make war with any deceitful machines or with poisoned weapons or with cannon, guns or any other kind of firearms.' This passage, it might be said, has been adduced in all seriousness as the reason for the non-employment of firearms in Britain during the times of the Druids, assuming thereby some mysterious and arcane connection between the Druids and the mystics of India at a time when India's existence was unknown to the people of Western Europe.

This belief in the Gentoo Code is of long standing; it appears to have arisen in 1773, when a first translation of the Code, certified to by no less a personage than Warren Hastings, quoted this passage and, further, used it to substantiate a doubtful claim that Alexander the Great had been confronted with firearms in India.

But in 1773 the translation of Sanskrit, the language of the Code, was an inexact science, and a certain amount of translator's licence had crept in. Later generations of philologists discovered that Sanskrit was by no means as simple as had first been thought, with eight grammatical cases and a complex code of rules for the formation of compound words. A much later retranslation of this same passage, in the light of improved knowledge, changed the sense completely: 'The King shall not slay his enemies . . . with deceitful or barbed or poisoned weapons, nor with any having a blade made hot by fire or tipped with burning materials.'

Most, if not all, of the claims for gunpowder's great antiquity can be dismissed in similar fashion. As long ago as 1866, one Lieutenant Henry Brackenbury—much later to be the Rt Hon. General Sir Henry, GCB, KCSI and President of the Ordnance Board—wrote a paper in the *Proceedings of the Royal Artillery Institute* entitled 'Ancient Cannon in Europe' in which he demonstrated the peculiar problems faced by historians in this particular field. He had discovered a reference in a highly regarded work which

BATAILLE DE MARIGNAN (Livrée en 1515)

Mémorial de l'Artillerie N° 6.

Pl. 1.

BATAILLE DE CERISOLLES (Livrée le 14 Avril 1544)

Pl. 2.

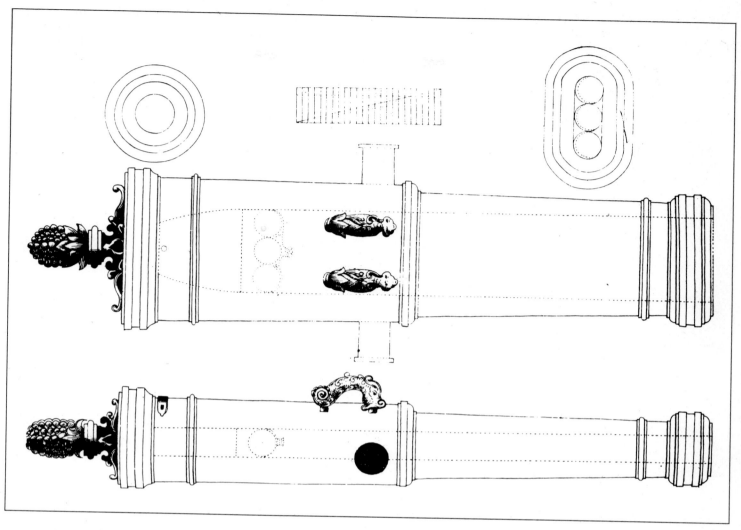

(*Above*) A 3-ball gun, proposed and made in the seventeenth century, but which proved to be impractical.

(*Left*) A print of the Battle of Marignan in 1515, between Swiss (on right) troops and the French. The cavalry charge, followed by infantry, is passing the guns. (*Royal Artillery Institution*)

(*Left*) The Battle of Cerisolles, 14 April 1544, between the forces of the Duc d'Enghien and those of Charles V. Pikemen are at grips, following the cavalry charge: the guns have served their purpose by this time, and are no longer manned. (*Royal Artillery Institution*)

mentioned 'two gun-founders' in a document from the reign of Edward II–well before gun-founding was generally thought to have begun. Brackenbury traced this reference to its source and discovered that the document in question had been misdated by some early archivist and actually referred to events in the time of Elizabeth I.

Nevertheless, the belief still persisted that gunpowder had been known in China or India, or possibly Arabia, or maybe Atlantis, many years before its appearance in Western Europe; even Biblical passages were quoted in attempts to argue such antiquity. The principal stumbling-block was the use of the word 'artillery' which, in various forms, appeared in many early documents and was invariably taken to mean cannon; whereas, in fact, 'artillery' was a general term covering ballistae, catapults and any other sort of missile-throwing engine. One might equally well deduce the existence of the internal combustion engine in the seventeenth century from the use of the word 'car', which at that time denoted nothing more significant than a light wheeled vehicle. As with 'car', so with 'artillery'; the new developments of automobiles and cannon merely took over the use of the accepted word for the general class of device. Since the records of the Middle Ages were less specific, the change of meaning which gradually took place went unremarked and thus led to confusion in later years.

Towards the end of the nineteenth century, due to the sudden revival of interest in artillery at that time, there was an equal rise in interest regarding its history and development, particularly in England, and a Royal Artillery officer, Lieutenant-Colonel Henry W L Hime, began writing a number of papers for publication by various learned societies. He followed this up by a long period of research into archives all over Europe in an attempt to solve this question of the inventor of gunpowder once and for all. One of the sources he checked was a text by Roger Bacon (1214–94) entitled 'De Mirabili Potestate Artis et Naturae' (On the Marvellous Power of Art and Nature), written in 1242. Bacon's name had frequently been advanced as a man who, if he did not actually invent gunpowder, appeared to have had a hand in its acceptance, but nothing in his writings seemed to give support to the claims on his behalf.

Among the various philosophical observations in Bacon's essay, Hime was struck by the incongruity of a string of meaningless words which appeared in the text just as Bacon seemed to be

warming to the subject of explosives. After much examination and thought, Hime hit upon the solution: the words were an anagram. After redistributing the letters and adding punctuation, the formula for gunpowder stood revealed: 'But of saltpetre take 7 parts, 5 of young hazel twig and 5 of sulphur and so thou wilt call up thunder and destruction if thou know the art.'

The question might naturally be asked, 'Why the anagram? Why not just put it down plainly so that all the world could see and give Bacon the credit?' The answer lies in the politics of the times. Bacon was a Franciscan friar; in 1139 the Second Council of the Lateran laid under anathema any person who made fiery compositions for military purposes. Had Bacon claimed to have invented gunpowder in plain language, or even alluded to its composition, he would have flouted this decree and probably hazarded his life. As it was, his outspokenness on many subjects and his other references to incendiary compositions in other writings led to a suspicion that he was a practitioner of the Black Art, and in 1257 Bonaventura, General of the Franciscan Order, interdicted Bacon's lectures at Oxford and ordered him into cloisters in Paris, where he remained incommunicado for the next ten years. Then, in 1266, the Pope requested him to prepare a number of treatises on the state of current scientific knowledge; these he wrote and sent to Rome, and he was allowed to return to his teaching at Oxford in 1268.

Among the papers written for the Pope was one entitled 'Opus Tertius', and in 1909 Professor Pierre Duhem of Bordeaux University came across a short manuscript in the Bibliothèque National in Paris which, on examination, proved to be a part of the 'Opus Tertius' and contained the following passage:

'From the flaming and flashing of certain igneous mixtures and the terror inspired by their noise, wonderful consequences ensue which no-one can guard against or endure. As a simple example may be mentioned the noise and flame generated by the powder, known in divers places, composed of saltpetre, charcoal and sulphur. When a quantity of this powder, no bigger than a man's finger, be wrapped up in a piece of parchment and ignited, it explodes with a blinding flash and a stunning noise. If a larger quantity were used, or the case were made of some more solid material, the explosion would be much more violent and the flash and noise altogether unbearable . . . These compositions can be used at any distance we please, so that the operators escape all hurt from them, while those against whom they are employed are suddenly filled with confusion.'

It will be noted that Bacon makes no claim to having discovered gunpowder in this, the first known plain-language reference to the substance. Instead he refers to its use in 'divers places', implying that such a composition was common knowledge and he was merely reporting a well-known fact of life.

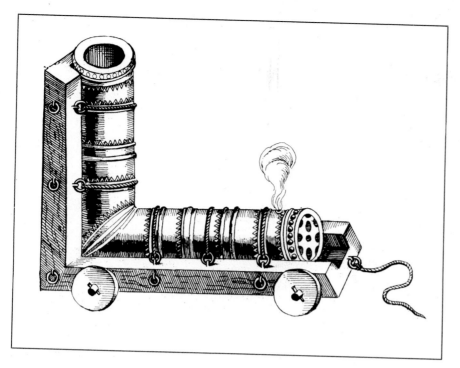

An optimistic design for a mortar, from *Artillery* by Uffano, published in Zutphen in 1621. Apparently the vertical section is the barrel and the horizontal section the chamber, but it seems unlikely to have been successful in practice. (*Royal Artillery Institution*)

The Millimete Gun, the first recording of an artillery piece.

Assuming then that powder was known as early as 1242, and was, on the evidence of Bacon's later work, common knowledge 'in divers places' by 1266, the next question to arise is that of who first used the substance to propel a missile and when and where?

Numerous spurious claims stand to be demolished here. Probably the most persistent story is that of Black Berthold, or Berthold Schwarz, the Mysterious Monk of Freiburg-im-Breisgau. An engraving by Custo, dated 1643, shows him in his laboratory launching the cover from an apothecary's mortar by means of a powder explosion; it is not clear whether this picture gave rise to the legend or vice versa, but the story goes that Schwarz, experimenting with powder, ignited a charge in a mortar and blew off the cover. From this accident he developed the idea of confining the powder in a cylinder and using

the explosion to propel a missile. This charming legend fails to withstand critical examination: the Custo engraving claims that Schwarz invented firearms in 1380, while other accounts offer 1354 as the critical year. In either case it can be shown that cannon were in use well before 1354, so that Schwarz's claim cannot be admitted. Moreover, recent research has led to considerable doubt about his existence at all, outside the realms of legend.

One claim which was given serious consideration for many years, and still is in some quarters, and which reinforced the Schwarz legend to some degree, was an entry found in the archives of the city of Ghent–'Memorialbuch der Stadt Ghent'–a sort of daily record of events–in which an entry dated 1313 stated 'in this year the use of "bussen" was discovered in Germany by a monk'. The word 'bussen' approximating to 'buchsen' was taken as meaning cannon. A further entry for 1314 observed that 'bussen and krayk (powder) were despatched to England'. However, later researches and re-examination showed that these entries had been added several years–if not centuries–after the claimed dates, and their validity can no longer be considered, in the absence of any corroborative testimony.

The earliest mention of artillery worth considering is probably the statement in Grafton's *Chronicles* that in 1267, during the rebellion of the Duke of Gloucester, Henry III approached London with his army 'making daily assaults, when guns and other ordnance were shot into the city'. There is, though, no other evidence to support this statement, and it would seem likely that Grafton fell into the 'artillery' pit and that Henry was, in fact, using the conventional type of throwing engines. Similarly Grafton refers to an incident in 1322 when Edward II fought the Battle of Leylade, in Northamptonshire, 'wherein he lost all his ordnance which were conveyed into Scotland'. This, too, is uncorroborated and probably refers to engines.

The first unquestionable reference is in 1326,

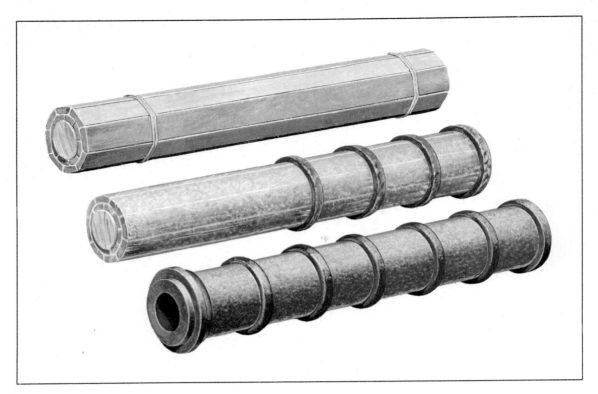

The early cannon were built by assembling longitudinal bars of metal and binding with hoops, the technique being based on the system used in cooperage.

when a Florentine document authorized the manufacture of brass cannon and iron balls 'for the defence of the commune, camps and territory of Florence'. Since metalworking was rather more advanced in Florence than elsewhere at that time, it is unlikely that cannon were in use at the affair at Leylade.

The year 1338 is the earliest date at which reliable testimony shows evidence of the existence of cannon in England and France. For the English part the evidence is an indenture between John Starlyng, former Clerk of the King's Ships, and Helmyng Leget, Keeper of the King's Ships, dated 22 June 1338, in which 'The said John doth deliver to the said Helmyng' in a ship *Bernard de la Tour*, 'ij canons de ferr' and on a barge *La Marie de Tour*, 'un canon de ferr ove II chambers un autre de bras ove un chamber'.

Grafton, although not entirely reliable on this subject, as we have seen, tends to corroborate this dating, since he refers to two of the King's ships, *Edward* and *Christopher* being lost to the French while escorting the English Army to Flanders in the same year, and he refers to the use of 'guns, bows and arbalests' in the action. (It should be noted here that 'de la Tour'–'of the Tower'–was the appelation given to all the King's ships at that time in order to identify them, much as HMS is used today.)

The French evidence is a document relating to the provision of a fleet fitting out at Harfleur for an attack on Southampton. It refers to the marine arsenal at Rouen, and mentions an iron cannon provided with 48 bolts made of iron and feathered, together with saltpetre and sulphur for making the necessary powder.

The year 1340 brings more artillery into the

records. Froissart first mentions the subject in a reference to the Siege of Quesnoy by the French, while the accounts for the city of Lille for that year show that payment was made to 'Jehan Piet de Fur, pour III tuaiux de tonnaire at pour cent garros, VI livres XVI sous'–'for three tubes of thunder and for 100 arrows . . .'

From then on, the references abound; in an account of the bailiffs of St Omer in 1342, printed in Napoleon's *Artillery* is an inventory of the castle

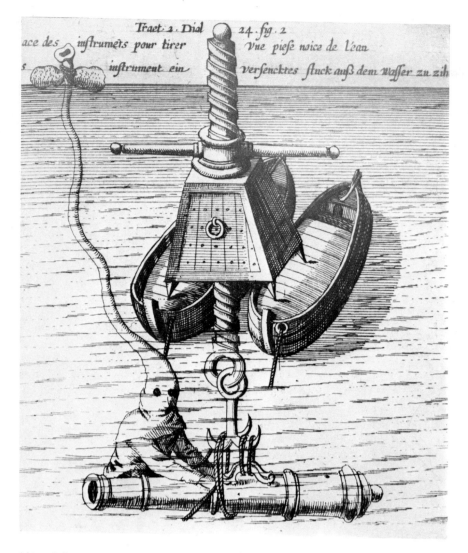

(*Above*) From Uffano's book of 1621, a sunken cannon being raised under the supervision of a diver. (*Royal Artillery Institution*)

(*Opposite, below*) A cannon dating from the end of the fourteenth century, carried on a primitive form of pivot mounting and probably intended as a defensive weapon for fortresses.

of Rihoult in Artois, including 400 shafts of arrows 'pour le canon' at 5 sous per hundred, winged with brass and bound with leather to touch the bore. The cannon in question had a separate chamber-piece held in place by a wedge. Another document from the Imperial Library in Paris, also quoted by Napoleon, showed that other projectiles had entered use by this time: a receipt given to one 'Ramundus Arquiero, artillerist to the French King at Toulouse' for sundry materials of war included two iron cannon, 200 lead bullets, 8 lb of powder and 200 wedges furnished with leather for the cannon.

In summary then, and on the evidence available, it seems that all that can be said with certainty is that gunpowder became known in Western Europe in about 1245 and was first used as a simple pyrotechnic. At some date between 1320 and 1326, some unknown experimenter applied the explosive force to the task of propelling a missile, and by 1340 the cannon had arrived on the battlefield.

Fortunately, when it comes to the question of what the early cannon looked like, we are on rather firmer ground, since there is evidence; evidence which at the same time reinforces our estimate of the date of its introduction. In 1325 Walter de

Millimete wrote a manuscript entitled 'De Officiis Regnum' and in this is an illustration depicting a cannon being fired. The picture is not explained by any text, but none the less it is possible to determine one or two salient features from the artist's impression. The first noticeable thing is the shape of the weapon, with bulbous breech and flared muzzle. This would seem to indicate an early appreciation of the need to reinforce the gun chamber as being the area of greatest pressure, and it is interesting to compare this shape with that of the American Parrot gun of the 1860s. The gunner appears to be lighting the charge through a vent, by means of a hot iron, the common ignition system of the early years, and the projectile, emerging from the muzzle, reveals itself as an arrow.

We have already seen that arrows were commonly referred to as cannon projectiles in early records, and the assumption can be made that since the soldiers of the day were familiar with arrows, it was a natural step for them to utilize this known form of projectile in the first cannons. In order to centralize the shaft in the bore and to seal the powder gases behind the arrow, a binding of leather was wrapped tightly round the centre of the shaft so as to make a tight fit in the bore. Again a remarkable resemblance can be seen between this, the first recorded gun-launched missile, and the Peenemunde Arrow Shell developed in Germany during the Second World War.

A certain amount of artist's licence seems to have crept in where the mounting of the gun is concerned. Allowing for the primitive perspective, it appears to be no more than a four-legged table or trestle upon which the cannon barrel rests. Were this the case, the intrepid gunner would doubtless have been knocked flat on his back by the recoiling cannon as it shot backwards off the table-top. However, the representation is close enough to the truth as we know it today; the original mountings were little more than baulks of timber grooved to form a bed for the cannon which was retained in place by iron bands and wooden wedges. This platform was either left free to recoil along the ground or anchored by wooden stakes. This system would suffice for the early weapons; the 'Millimete' gun appears to be about three feet long, with a bore probably in the region of two inches in diameter. Allowing for the fact that the gunpowder of the time was relatively weak, and that the projectile was light, the recoil force would be safely restrained by such simple measures.

At this stage of development, of course, there was no question of such refinements as sights or methods of applying elevation; it was remarkable enough that the weapon worked at all, and the very fact of its discharge was often enough to weigh the scales of battle in its owner's favour, without having to worry about actually aiming it at anybody in particular; if the missile struck someone, it was a bonus. One account of the Battle of Crécy in 1346, by Mezeray, says that

King Edward 'struck terror into the French Army with five or six pieces of cannon, it being the first time they had seen such thundering machines'.

Unfortunately the Millimete illustration gives no certain indication of the method of construction of the cannon, but it was probably cast of bronze or bell-metal; the colouring of the drawing appears to have been carefully done, and the dull yellow of the cannon argues against cast iron as a material. Moreover the casting of iron is not generally credited at such an early date, though the technique of casting bronze and other metals of low melting-point was well known, as for example, in bell-founding.

This type of weapon, known variously as 'pot-

Once this system of construction was mastered, the size of the cannon was no longer restricted by the imperfect casting techniques or the expense of castable metal, and from the small-bore light guns of the first days, calibres soon increased.

At the Siege of Odruik by the Duke of Burgundy in 1377, cannon throwing stone balls of 200 lb were employed, and in the same year there is a record of the construction of a cannon for the same Duke to fire a 450-lb shot, which roughly equates to a calibre of 22 inches. One is inclined to wonder why, with such massive ordnance available, sieges—of which there were plenty in those days—still took so long. It would be reasonable to expect that a few shots from a 22-inch gun would

de-fer', 'vasii' or 'sclopi' appear to have launched the cannon on its long career, but within a few years its defects led to its abandonment. The soft and relatively expensive bronze soon gave way to iron—hence 'pot-de-fer'—but the casting process was involved and only lent itself to the production of small weapons by highly skilled artisans, and a simpler and cheaper method, which would allow larger weapons to be made, was soon adopted.

It is not surprising that the similarity of form between a cylindrical hollow cannon and a cylindrical hollow barrel led to exploration of the cooper's technique as a method of constructing guns, and by 1360 the standard method of gun-building was to fabricate the body of the weapon from wrought-iron strips placed lengthwise on a mandrel and hammered so as to weld them together. These were then reinforced by circumferential hoops, a system of fabrication which led to the adoption of the name 'barrel' for the final product. In order to strengthen and preserve the iron, this basic barrel was tightly wrapped with rope and finally encased in leather, to prevent the rope rotting and the metal rusting.

soon breach any contemporary work and terminate the investment. But here we find the first appearance of the influence of ammunition on weapon performance and thus on the course of the battles.

The contemporary gunpowder was weak stuff: Bacon's formula was for 41 per cent saltpetre, 29.5 per cent sulphur and 29.5 per cent charcoal, and no doubt the materials of his day were to no great degree of purity. Modern gunpowder, on the other hand, uses the proportions 75:10:15, together with insistence on a high grade of purity throughout. Moreover, the early powder was made by grinding the dry ingredients into fine powder in a mortar and then mixing them by hand; this fine mixture, when loaded into a gun chamber, consolidated so that ignition was difficult, the flame being unable to penetrate the fine mass very quickly. This led to some uncertainty of action, much of the powder being unconsumed by the time the shot left the muzzle, and consequently a portion of the charge was ejected unburned—a serious defect in those days, since the powder was the most expensive item. While iron in 1375 cost

A specimen of a peterara, showing the system of breech-loading by using a removable breech-piece.

A 9½-inch bronze mortar captured by the British in 1838 and probably dating from the time of Tippoo Sahib in the late eighteenth century.

The 'Tsar Puschka' or Great Gun of Moscow on a travelling carriage. Cast in Moscow in 1586 by Andrea Inchochov, it bears a likeness of the Tsar Theodore.

'Mons Meg', the Edinburgh cannon. Of 19½-inch calibre and weighing five tons, this early example of heavy ordnance dates from the latter part of the fifteenth century and is believed to have been made in Flanders.

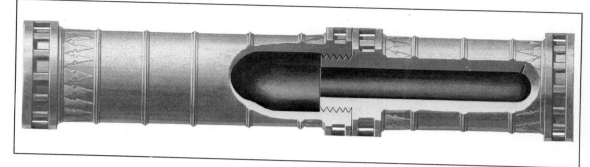

The Great Gun of
Mohammed II, cast in 1464.
It formed part of the
armament of the Castle of
Asia at the mouth of the
Dardanelles and was
presented to Queen
Victoria by Sultan Abdul
Aziz Khan in 1866.

$2\frac{1}{2}$ pence per lb (at modern equivalent prices) and
lead 5 pence, gunpowder cost the staggering sum
of 60 pence per lb, largely due to the scarcity of
saltpetre.

The large battering guns used stone shot and,
of course, their effect depended upon the relative
hardness of the stone used for the shot and that
used for the defensive work which was being battered. This use of stone shot also stemmed from
the weaknesses of powder and cannon. Supposing
the guns to have been strong enough, the poor
powder would have delivered insufficient velocity
to an iron ball of 22-inch calibre (which would
weigh about 1,480 lb instead of the 450 lb of a
stone ball). But the structure of the gun alone was
sufficient to rule out the use of an iron ball of
such weight; even weak powder would have developed a high pressure within the gun chamber
before it overcame the inertia of the ball, and this
high pressure would inevitably have burst the
weapon asunder.

At this time, the relative immobility of ordnance considerably limited its tactical use. While
the cannon remained a simple tube lashed to a
wooden bed, transportable only by cart or waggon
and emplaced by considerable exertion, it was obviously not a weapon which suited the rapid
movements of an encounter battle. To use a phrase
from later years, it was 'artillery of position', em-

ployed only in set-piece battles or sieges where
there was ample time in which to deploy it. At first
this was no deprivation, for the engines of earlier
days had been even more ponderous. The idea of
light firepower to accompany the foot soldiers was
not even considered. But towards the end of the
fourteenth century comes the first report of
wheeled artillery capable of moving with the
army on foot and designed with the intention of
employment in support of infantry. Froissart reports that in 1382 the men of Ghent, setting forth
to have at the citizenry of Bruges, took with them
a number of 'ribauldequins'–light, two-wheeled
carts mounting a number of small-calibre cannon
and protected from attack by iron spikes projecting in front of the cannon muzzles. These weapons
were placed at the forefront of the army when
drawn up for battle in order to protect the main
body from sudden attacks. Unfortunately they
were not capable of being reloaded rapidly, and
their first shot in action was likely to be their last,
since the tide of action would have rolled past
them before they could be readied for a second
volley. At the Siege of Oudenarde, also in 1382,
the besieging Flemings were taken in the rear by
a relieving French army under Charles IV. Their
ribauldequins were redeployed to meet this new
threat and, on the advance of the French, were
discharged with such an effect as to make the

(Below) An engraving of
1697 depicting a cannon
foundry. The illustration
below right and those on
the following two pages
are engravings showing the
stages in the manufacture
of the cannon.

French Army 'recoil one pace and a half'. But once the guns had spoken and the French had recovered their composure, they swept into the Fleming ranks and signally defeated them, leaving 15,000 Fleming dead on the field.

The slow process of loading, by ladling a measure of powder down the bore and following it up with the ball, priming the vent with powder and finally touching off the gun, interspersed with applications of a wet brush to remove the crusted powder-fouling and extinguish any lingering spark, plus the hard labour of dragging the gun back to the firing position from wherever it had come to rest after firing, led to early attempts to improve matters. We have already noted in passing that as early as 1338 there is a reference to 'une canon ove II chambers' and in 1342 one to cannon with a separate chamber held in place by a wedge.

These weapons were no more than a cannon barrel open at both ends, laid in a wooden bed. The chamber and breech end was a separate item which was laid in place behind the rear end of the barrel and held there by driving wedges between the base of the chamber and an upstanding block of wood forming the rear end of the bed. By providing the gun with a number of spare chambers and loading each one before the start of a battle, it would be possible to fire a number of shots in fairly rapid succession by removing the fired chamber and replacing it with a loaded one. Doubtless a slow but steady fire could then have been kept up by reloading the chambers as they were removed.

This design was later improved into the 'peterara' in which the gun was made in the normal way, after which the breech was cut away to allow a

The sequence of pictures on these two pages shows stages in the manufacture of a cannon from engravings of 1697. See also the pictures on page 19.

removable chamber to be dropped in and wedged. Another form of construction was to prepare an open-ended barrel with metal arms extending to the rear to hold the removable chamber. These systems of construction made for a more certain connection between chamber and gun, since the chamber was now held between two faces of the same piece of metal, and hammering home the locking wedges would no longer be likely to force the barrel forward on its wooden bed.

Although on the face of it this sounds an effective and simple system—after all, it is precisely that used by the modern revolver in its basic principle—it was grossly inefficient. The unknown innovator had little comprehension of the magnitude of the forces involved in the explosion or of the speed of its action, and the imperfect fitting of chamber to breech—for the tools and ability to

make perfectly mating faces to a fine tolerance were several hundred years in the future—led to severe gas leakage. The slow-burning powder compounded the mischief, developing its power so slowly as to allow considerable leakage before the projectile was ejected from the muzzle, and this leakage of gas eroded the joint faces, such as they were. This in turn led to greater leaks and faster erosion until eventually the weapon became positively hazardous to fire and was discarded. There are numerous peterarae in museums, a notably fine collection being that of the Museu Militar in Lisbon, and in spite of the ravages of time it is quite often possible to discern marks of erosive wear at the faces of chamber and barrel.

Another system of breech-loading which probably dates from the early years of the sixteenth century is exhibited by a gun now in the Royal

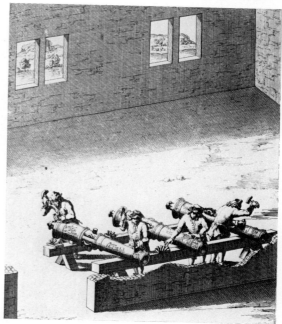

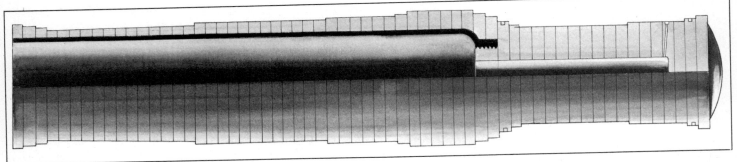

'Dulle Griete' (or 'Mad Margaret'), a 25-inch calibre monster cast in Flanders about 1430. It bears the arms of the Duke of Burgundy.

Artillery Museum at the Rotunda at Woolwich. In this the chamber-piece fits into an enlarged section of the gun barrel and is retained there by an iron bar passing through holes in both the gun body and the chamber-piece. This is, in theory, a slightly more efficient design, but again the imperfections of construction would soon lead to gas leakage and wear. The origin of this specimen is not known, but a similar weapon of German design is shown on page 27.

Having seen the beginnings of artillery, it might be as well to pause at this stage–say at about AD 1425–and ask what effect this new introduction had had upon the warfare of the time. The short answer, remarkably enough, is 'Very little'. Had these early guns produced a result at all in keeping with the expense and difficulty of constructing them and the labour of moving them, artillery might have received more attention than it in fact did. At this period cavalry had recently adopted plate-armour, proof against swords and lances, which made the contemporary cavalry charge wellnigh irresistible. Now, if the early cannon had been able to make a significant impression by delivering a massive counterblow to the mounted arm, it would have been of considerable importance to the English Army of the period, since the vital component of this force was its archers and foot soldiers, whereas the French Army of the day was predominantly a cavalry army.

But the weak guns, the slow rate of fire, and the difficulty of moving the guns from place to place prevented any sweeping changes in the technique of war. None the less the effect was sufficiently marked to ensure the adoption of artillery by every army. For what the cannon lacked in one sphere it made up in another: though it killed few, it terrified many, and its effect on morale was considerable. Hitherto warfare had been a simple affair of strength versus strength; he who brought the strongest force on to the field usually departed as the victor. Brute force was the sole criterion, and the number of dead of either side left on the field was the gauge of victory or defeat.

The arrival of gunpowder changed this, since it placed an equal force in the hands of the weak or the strong, and it was the intellect which guided the application of the force which would eventually prove victorious. The gun was feeble, inaccurate and slow, true, but when the missile was launched, death inevitably followed. Courage, armour, numbers, rank were no proof against it, and it took

effect at ranges hitherto undreamed of in combat. Yet it was not the destruction it actually wrought that was its prime strength; it was the destruction it threatened. The flame, smoke and explosion were a new and terrifying addition to the hazards of war; and when the flying ball cut down a splendid knight, in spite of his armour, his horse and his skill at arms, who knew what the next might not do? It was this levelling, this threat, this promise of sudden and complete destruction, which artillery brought to the field and which, gradually gave it prominence.

In the earliest days, though, much of this lay ahead. At this time its influence was almost entirely due to its surprise and effect on morale, since its physical effectiveness was severely limited. Many years were to elapse before the handling of artillery, so that it could exert a decisive influence on the battle, was explored, let alone understood.

Another factor militating against artillery's advancement in those early days was the amount of expensive material necessary to equip a force of artillery, an amount far exceeding the difficulty and cost of equipping an equivalent force of cavalry or infantry. In the fourteenth century every peasant was familiar with the use of bow, pike and sword; every gentleman owned horses and armour. At a moment's notice an army could be assembled from these elements, an army the equal of any likely enemy force. But artillery could not be produced in similar fashion; there are one or two early records which indicate that private individuals owned cannon and hired their expertise and equipment to the King in time of need, but this was exceptional. Cannon and powder were expensive items, and they also demanded a retinue of expensive attendant specialists, and neither the equipment nor its practitioners could be created out of thin air when war threatened–though governments, even to the present day, have persisted in trying to do this very thing.

In consequence artillery had to be produced in time of peace and both it and its train of attendants retained against the time of their need in war. Since the Middle Ages had but rudimentary ideas of economics or fiscal science, few kings possessed the financial resources to maintain such expensive luxuries in numbers which would have made an appreciable impression in war. For example, Edward III in 1360 possessed but four guns and 16½ lb of powder, and at Calais, the English Army's base of supply for the Hundred Years

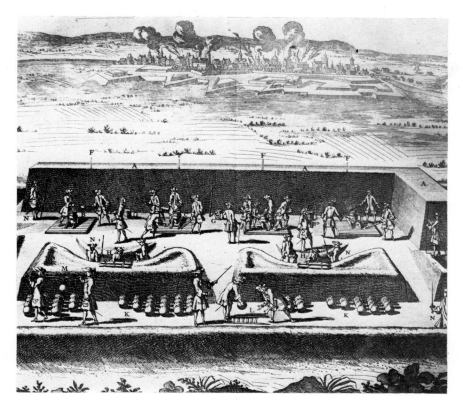

A mortar battery of the early eighteenth century takes up position in front of a besieged fort. (*Royal Artillery Institution*)

War, an inventory in 1370 gave fifteen guns and 84 lb of powder as the Army's entire resources.

Yet another factor preventing the early growth of artillery was a more curious trend: the rise, during the fourteenth and fifteenth centuries, of the mercenary soldier. These were mercenary in every sense: not only were they paid for their services, but they entered battle with the intention of surviving to show a profit. A dead enemy was worth only the price of his armour and possessions, but a live captive was a potential source of ransom money. Obviously these gentry were opposed to the employment of such lethal instruments as cannon, which were just as likely to kill a knight of considerable financial potential as a worthless bowman. Where mercenaries were involved warfare became a lucrative but otherwise slightly ridiculous activity; at the Battle of Zagonara in 1423 only three men were killed, and they were suffocated by falling into soft mud from which the weight of their armour prevented their rising. And in a battle between Neapolitan and Papal troops in 1486, which lasted the entire day, no one was killed or even wounded.

It was only in the siege that artillery had a chance to display its destructive ability. Here, in actions which were as formal and long-drawn-out as a chess game, there was time to assemble the ordnance, position it for best effect, and open fire in the knowledge that it would be possible to fire more than one shot before the affair was over. In 1380 the Venetians besieged Brondolo and opened fire with two cannon of considerable size, so great that one stone ball demolished a large part of the wall of the Campanile and killed 22 men at a single stroke. And, of course, the

famous Siege of Constantinople in 1451 witnessed the astonishing spectacle of the besiegers actually sitting down outside the walls and casting their cannon on the spot. This became a standard practice with the Turks; Mohammed ordered the casting of 16 guns before the walls of Rhodes in 1480. These massive pieces were over 18 feet long and threw stones of 25 inches in diameter 'which flying through the air by force of powder fell into the city and, lighting upon houses, broke through the roofs, made their way through the several stories, and crushed to pieces all they fell upon; nobody was safe from them and it was this kind of attack that gave the greatest terror to the Rhodians'.

An interesting feature of these huge Turkish cannon is that they remained serviceable ordnance for almost 400 years. After their use in various battles they were always brought back to Constantinople and mounted alongside the Dardanelles, and in 1807 a British squadron attempting a passage was bombarded by these antiques. For all their great age and immobility, they could still draw blood. HMS *Standard* was struck by a 770-lb stone ball 'which killed four men and led to a succession of disasters by which four more lost their lives and four were injured'. HMS *Active* was struck by an 800-lb stone 'which made so large a hole in the side that the captain, looking over to see what was the matter, saw two of his crew thrusting their heads through it at the same moment', while HMS *Hercules* suffered ten men killed and ten wounded.

Such monster cannon were more common in the East than in the West, doubtless because the mighty potentates of the East could better afford them. The 'Mukh-el-Maidan' (Master of the Plain), the Great Gun of Beejapore, was cast at Ahmednuggar in 1548 under the superintendence of a Turkish expert. Of 28·5-inch calibre, it fired a 1,000-lb stone ball by means of an 80-lb powder charge. It appears to have been fired last at some time in the eighteenth century, on the occasion of a state visit from a neighbouring rajah, when, according to legend, all the pregnant women within earshot miscarried.

Another Indian gun, the 'Dhool Dhanee' (The Scatterer) of Agra had a 23·5-inch bore, but was broken up and sold for its metal in 1832. Further west was the 'Tsar Puschka', the Great Gun of Moscow cast in 1586. Until 1944 this was the largest calibre gun ever made, with a 36-inch bore. The stone ball would have weighed something in the region of 2,400 lb, though there is no record of the gun ever having been fired.

Ordnance such as these could wreak tremendous damage on walls and buildings erected before such weapons were ever imagined, and thus in sieges, if in no other type of battle, artillery could produce a material effect fully as formidable as its moral effect. But the siege was to become a less common event as warfare gained mobility; and if artillery was to keep its place, it, too, had to become mobile.

CHAPTER 2

From Art to Science

Les proportions de la petite Piece de Canon
de fonte de deux livres de balles, sont marqué
et cottées tant sur la Piece ou sont les orne-
mens que sur la coupe d'icelle qui est a costé

*F*or about 250 years the artillery remained a minor arm; indeed, it very nearly disappeared from the battlefield altogether. The simple character of battles and the almost total absence of tactical movement tended to restrict the activity of the guns. The normal course of battle in those days was for the opposing armies to be drawn up opposite each other, with the infantry in the centre, the cavalry on the wings, and such cannon as were available distributed across the front. A volley from the cannon opened the proceedings, after which the cavalry charged. The infantry whose cavalry were defeated in the charge fell back pursued by their opponents and the immobile guns fell into the victor's hands. Since the armies were, for the most part, composed of untrained levies lacking in drill or discipline, little in the way of advanced tactical movement could be expected.

None the less a certain amount of technical improvement took place during these fallow years. The major step was the improvement of gunpowder. The original powder, known as 'serpentine', had certain defects; as well as its fineness, upon which we have already remarked, the mixture had a tendency to separate back into its constituent parts when subjected to vibration, such as when being carried in casks in an unsprung cart across indifferent tracks. The heavier sulphur and saltpetre would descend to the bottom of the mixture, leaving almost pure charcoal at the top. This meant that the gunner had to remix the powder on arrival at the scene of action, and with gunpowder's great susceptibility to ignition by friction, this was a dangerous practice. The alternative, often employed, was to carry the three ingredients separately and mix them as needed, a no less hazardous procedure.

The improvement, like the discovery of gunpowder in the first place, is difficult to pin down to a particular date or person, but it appears to have been developed in France and the first mention of it is in 1429; the substance of the improvement lay in that the powder was mixed in a wet state, which made for better incorporation and less liability to accidental explosion, after which it was allowed to dry into a cake which was subsequently broken up and the resulting grains passed through sieves in order to regulate the size. This 'corned' powder showed great advantages over serpentine; the granular form allowed flame to penetrate the charge more easily, thus giving faster and more efficient ignition, so much so that corned powder was estimated to be about one-third more powerful than the equivalent amount of serpentine. It

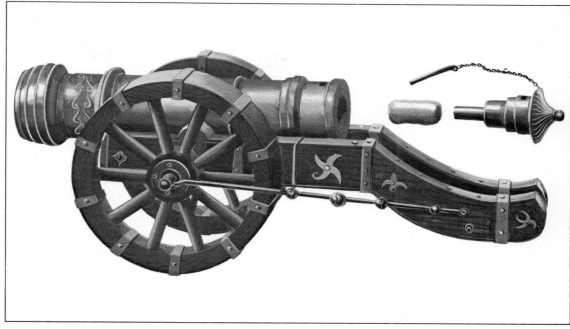

(*Left*) Frontispiece from *Memoirs d'Artillery* printed in Paris in 1745. On the left a mortar is being prepared for firing; in the centre a gunner takes aim, while to his right engineers are constructing an embrasure. In the foreground there appears to be a class in elementary ballistics in progress. (*Royal Artillery Institution*)

(*Right*) A German breech-loading gun of the sixteenth century. The breech plug was retained in place by an iron spike passed through a hole in the gun chase and engaging in slots cut in the body of the plug.

(*Left*) A *Canon pour tres coupes* in *Memoirs d'Artillery*, 1745. Unfortunately the accompanying text is not very explicit, and it is not at all certain whether the three barrels were to be fired separately or together. (*Royal Artillery Institution*)

(*Right*) Sixteenth-century German mortar from Seftenberg; the bomb rests on the chamber with its fuze against the propelling charge. A ringbolt on the other side of the bomb aided the process of loading.

(*Far right*) A seventeenth-century mortar quadrant, placed in the muzzle to determine the elevation and marked off in 'points'.

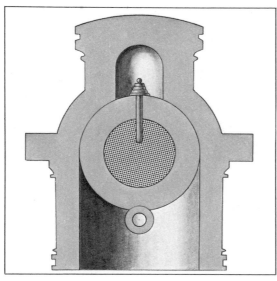

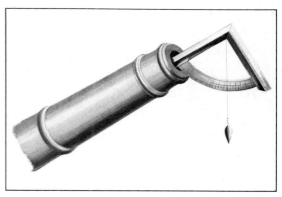

was also less susceptible to moisture, it no longer separated out during transportation, it demanded less care in ramming into the gun and it left less residue after firing.

Such is the way of improvements that, of course, corned powder was not without its own brand of problem. The two greatest drawbacks to the universal employment of the new powder were firstly its expense, since the involved method of manufacture greatly increased the already high price, and secondly its greater power which was frequently too much for the weak cannon of the day. Owing to these factors corned powder was slow to gain acceptance. But if it did nothing else it accelerated the development of cannon of a more robust nature than the type built up from bars and hoops. The technique of casting had now made progress, and in spite of its expense it was obvious that this form of construction was the only one which would produce practical weapons. Bronze guns were cast in France from about 1460 onwards, and when cast iron became more familiar, it too, was adopted.

The first iron cannon produced in England were cast at Buxted in Sussex, in 1543, by Ralph Hogge, Peter Baude and Peter Van Colin. The Sussex Weald at that time was the premier iron-working area of England, due to the proximity of iron ore and ample wood for producing the charcoal required for the furnaces. Previous to this some bronze cannon had been cast in England by one John Owen, though the location of his foundry is obscure. From this time forward the built-up gun disappeared from the scene, the greater strength of the cast weapons, which allowed corned powder to be used, ensuring their general adoption. However, cast iron, in spite of its economy, was some time in ousting bronze, and it was not until casting techniques improved and reliable weapons could be guaranteed that iron really became common. It is not without interest to see that bronze was never entirely discarded—bronze guns, albeit of modern design, were still in use by the Italian Army during the Second World War.

By this time cannon were taking on a variety of

shapes and sizes to suit their employment or the whims and fancies of their constructors or purchasers, and by 1574 there was a sufficient variety in use to allow certain titles to become standardized. Where some of these titles originated is obscure; they are largely based on actual or mythical birds, but one or two defy reasonable analysis:

The Robinet	1·5-inch calibre, 1-lb shot
The Falconet	2-inch calibre, 2-lb shot
The Falcon	2·5-inch calibre, 2·5-lb shot
The Minion	3·25-inch calibre, 4·5-lb shot
The Saker	3·5-inch calibre, 5-lb shot
The Demi-culverin	4·5-inch calibre, 9-lb shot
The Culverin	5·5-inch calibre, 18-lb shot
The Demi-cannon	6·5-inch calibre, 30-lb shot
The Extra Cannon	7-inch calibre, 42-lb shot
The Cannon	8-inch calibre, 60-lb shot

Another, later, list from the *Compleat Souldier* of 1628 shows some changes:

	Weight lb	Calibre in	Length ft	Shot lb
The Falconet	400	2·25	6	1·125
The Falcon	750	2·75	7	2·5
The Sakeret or Minion	1,100	3·25	8	4·75
The Saker Ordinary	1,900	3·75	9·5	6
The Demy Culvering	3,000	4·5	11	11·75
The Ordinary Culvering	4,300	5·25	12	16·25
The Culvering	4,600	5·5	13·25	19
The Demy Cannon	5,000	6	11	24·5
The Demy Cannon Ordinary	5,600	6·5	10·25	32
The Demy Cannon Eldest	6,000	6·75	11·25	36·5
The French Cannon	6,500	7·25	12	46·75
The Cannon Serpentine	7,000	7·5	11·5	52
The Cannon	8,000	8	12	64

At about this time, too, the first writings on artillery and gunnery made their appearance. The earliest book devoted to the subject was *La Nova Sciento Invento* by Nicolo Tartaglia, published in Venice in 1537, and it was closely followed by *Pyrotechnia*; *On the Art of Improving the Quality of Gunpowder and Casting Metal for Cannon* by Beringuccio. The first English work was Bourne's *Art of Shooting Great Ordnance* published in 1578.

Tartaglia's work, plus others he wrote in 1546, were translated into English in 1588 by Cyprian Lucar, who added an appendix, 'Shewing the Properties Office and Duties of a Gunner', and a few extracts from this are worth repeating:

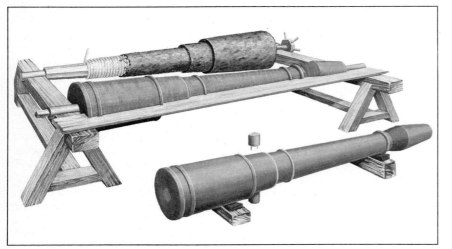

(*Above and right*) Before casting a cannon it was necessary to make a model from which the mould could be built; this shows a model in the processes of manufacture. A wooden core, wrapped with rope, was built up with clay to the general shape required; this was then turned to the correct form by a simple template; finally the trunnion pieces were added. The odd shape at the muzzle was to facilitate pouring the metal and also to provide waste metal into which the impurities would rise in casting.

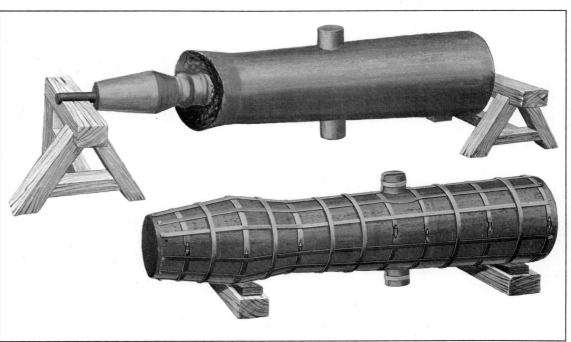

(*Right*) With the gun rough cast, the bore had to be accurately finished in this vertical boring machine.

(*Below*) Cannon balls were also moulded, in this simple former. The plug of metal left by the pouring vent was removed with a chisel and the cut surface filed smooth.

(*Bottom*) A drawing of about 1620 illustrating the ballistic beliefs of the time; that direct fire guns shoot in straight lines and mortars have a three-piece trajectory, the *motus violentus* under the effect of the powder, the *motus mixtus* where the effect of the powder is wearing off and that of gravity appearing, and the *motus naturalae* in which the projectile falls straight to earth.

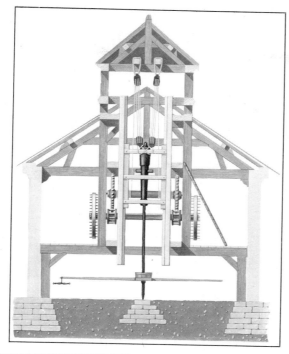

'A gunner ought to be sober, wakefull, lusty, hardy, patient, prudent and a quick-spirited man, he ought also to have good eyesight, a good judgement and perfect knowledge to select a convenient place in the day of service, to plant his Ordnance where he may do most hurt unto the enemies and be least annoyed by them . . .

'Also he ought to be no surfieter nor a great and sluggish sleeper, but he must govern himself at all times as a wise, modest, sober, honest and skilful man ought to, that through want of understanding he may never lose his credit nor an universal victory which oftentimes by the means of good Gunners well managing their pieces is gotten . . .

'Also a Gunner in time of service ought to forbid with meek and curteous speech all manner of persons other than his appointed assistants to come near his pieces, to the end that none of his pieces may be choked, poisoned or hurt, and he ought not for any prayers or reward lend any piece of his gunmatch to any other person because it may be hurtful to him in time of service to lack the same . . .

'Also if a Gunner charge his piece with cartredges he ought to sett them upright in a tubb or some other wooden vessel which (though it seem to stand in a place out of danger for fire) should never be uncovered for any longer time than while the same cartredges are taken out one by one to charge the piece . . .

'Also every gunner before he shoots should consider whether the air is thin and clear or close and thick, because a pellet will pass more easily through a thin and clear air . . .

'Also every Gunner ought to know that as it is a wholesome thing for him to drink and eat a little meat before he doth discharge any piece of ordnance, because the fume of saltpetre and brimstone will otherwise be hurtful to his brains, so it is very unwholesome for him to shoot any piece of ordnance while his stomach is full . . .'

Leaving aside the exhortations to be of upright character and abstemious habits, there are some significant points here; the mention of 'cartredges' for example, shows that by this time it was an accepted practice to weigh and bag gunpowder into cartridges for convenience in carriage and loading, instead of simply ladling loose powder down the bore. Furthermore his reference to 'thin and clear air' and 'close and thick air' shows some comprehension of the effects of atmospheric conditions on the flight of the projectile, a theory which was still looked on as something revolutionary in some quarters as late as the early years of the twentieth century. But Lucar's appreciation of this and other problems—his appendix also discussed the effects of shooting up- or downhill, shooting with one gun wheel higher than the other, and the problems of quartering or following winds—was obviously based on long experience and observation. He was not always prepared to advance theories as to *why* the piece 'shooteth

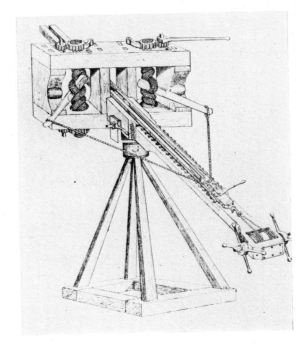

over the marke' or 'wyde of the marke'; he merely pointed out that it did so under certain circumstances and he urged practical gunners to avoid the circumstances in question. For the science of gunnery was as yet unborn, and ballistic theory of those days–what little there was–was highly erroneous. It was, for example, widely believed that the ball, on leaving the gun, travelled in a straight line in prolongation of the barrel's axis until it reached maximum range, whereupon it stopped and fell to the ground.

But Tartaglia (1500–57) was a professor of mathematics at Venice, and his *Three Books of Colloques concerning the Art of Shooting*, translated by Lucar and dedicated to Henry VIII were the first attempt to introduce some scientific method into the Black Art. Tartaglia was the first to point out that 'A peece of artillery cannot shoot one pace in a straight lyne' and that 'the more swift the pellet doth fly, the less crooked is its range' or, in modern parlance, the faster the velocity, the flatter the trajectory. He was also the first man to point out that the maximum range of a gun would be achieved at an elevation of 45 degrees, a statement which remained true until the twentieth century produced weapons capable of pitching their projectiles into the stratosphere. But the path of the innovator is ever hard, and Tartaglia's theories, while discussed by learned men of the day, were somewhat beyond Lucar's sober, wakeful and lusty gunners.

During the fifteenth and sixteenth centuries the handling of artillery made little progress, largely due to the difficulty of moving the weapons and the lack of a tactical doctrine suited to the guns. The first steps towards improvement were taken by Gustavus Adolphus (Gustav II of Sweden, 1594–1632). He appreciated the value of artillery correctly handled; doubtless others before him had come to the same conclusions, but Gustavus was in a position to do something about it. His first step was to draw a line dividing artillery into

(*Right*) An early 'war engine', in this case an extremely powerful crossbow, which was probably more effective and accurate than the early cannon.

(*Far right*) A catapult, illustrating the mechanical ingenuity which went into the final versions of these devices.

(*Opposite, centre*) A Swiss falconet, dating from 1672; the frail-looking carriage threatens instability, showing that even by this date the gun-makers were not yet certain of the best design.

(*Opposite, far right*) One of Gustavus's famous 'Leather Guns', showing the system of construction. A copper barrel was tightly bound with rope before shrinking on the outer leather cover.

(*Right*) An Italian gun from the fourteenth century. The carriage has progressed slightly from the wooden bed, being furnished with wheels and a rudimentary form of elevation apparatus.

(*Far right*) An old print showing a catapult in a defensive role, and an early application of indirect fire.

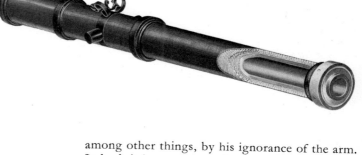

two branches, field and siege, classing everything about the 12-pounder as a siege gun. His next move was to introduce the celebrated 'leather guns'. These were not entirely of leather, of course, but were lightweight copper barrels securely bound with rope and leather to reinforce them. There is evidence to suggest that these guns were actually invented in Scotland and the idea taken to Gustavus by a Scottish mercenary; be that as it may, Gustavus was the first to put the idea to good use. Mounted on light two-wheeled carriages these weapons were more mobile than anything previously seen in battle, though they were a retrograde step as far as power and accuracy were concerned. None the less, what they lacked in range they more than made up for by the ease and speed with which they could be manœuvred.

Gustavus's final move was to combine the movement of cavalry and guns so as to take advantage of the mobility while bestowing covering fire on the movement of the mounted arm. These simple tactics, precursors of all 'fire and movement' theories, were first put to use at Dirschen and then at Leipzig in 1631. Here Gustavus won the battle by, among other things, his appreciation of the value of mobile artillery, and Tilley lost it,

among other things, by his ignorance of the arm. Indeed, it is not too much to say that Tilley's inexperience with artillery was fundamental to his defeat, since he made three gross blunders, each involving the guns: he positioned his guns behind his infantry, firing over their heads, so that any movement of the infantry would mask the guns; having done that, he *did* move the infantry and *did* mask his guns; and finally he allowed Gustavus to change front, a manœuvre which Gustavus covered with his leather guns, and as a result Tilley's guns were later captured and turned against him.

Unfortunately Gustavus was killed at Lutzen in 1632–another battle in which his handling of artillery had a decisive effect–and with his passing the brief period of enlightened artillery tactics passed with him. His successors did not possess his breadth of vision or his particular gift for handling guns, and within a few years artillery was once more in obscurity. In England the ability of the Navy ensured safety from invasion to such a degree that the equipment of the Army was grossly neglected. When the Civil War broke out the Parliamentary troops began by owning no

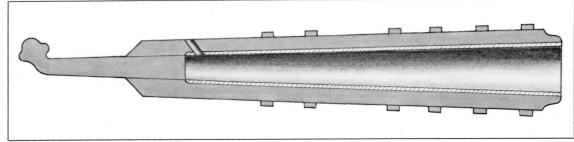

artillery at all, but the Royalists were so inept at handling what few guns they had that no advantage accrued from their possession. Macaulay points out that at Sedgemoor in 1685 'so defective were the appointments of the English army that there would have been much difficulty in dragging the great guns to the place where the battle was raging had not the Bishop of Winchester offered his coach horses and traces for the purpose'. Even when the guns reached the field there were insufficient gunners to operate them, and a Serjeant Weems of Dumbarton's Regiment took over the supervision and control of a number of the guns thereby earning himself a gratuity of £40 'for good service in the action at Sedgemoor in firing the great guns against the Rebels'. In spite of this want of efficiency the guns turned the scale and brought the battle to an end, the rebels being put to rout.

After the Restoration, artillery appears to have vanished from sight in England, for Macaulay tells that when William of Orange landed (1688) 'the apparatus he brought with him, though such had been in constant use on the continent, excited in our ancestors an admiration resembling that which the Indians of America felt for the Castilian harquebuses'. This 'apparatus' consisted of '21

huge brass cannon which were with difficulty tugged along by 16 carthorses each'.

One of the other causes of artillery's poor standing at the time was that the force rarely belonged body and soul to the Army. The problem of maintenance of such an expensive and technical force in peacetime, already touched upon, was still unsolved. A limited number of professional gunners were retained, together with a number of guns, and when war broke out this cadre was augmented by a scratch collection of labourers and drivers to serve under the gunners. A great difficulty lay in the fact that these reinforcements were hired civilians rather than soldiers, and when things got too hot for them, they frequently decamped, leaving guns and gunners to manage as best they could. Sooner or later this misfortune befell most armies, and sooner or later the fact was accepted that the expense of forming a permanent corps of artillery simply had to be borne. In this way the entire force, gunners, drivers, fireworkers, matrosses and other peculiar incumbents were subject to the same military discipline and imbued with the same martial spirit as the rest of the Army.

The War of the Spanish Succession (1702–13) shows some leanings towards a resurgence of flexible artillery employment which had been forgotten since Gustavus's time. Marlborough, to everyone's surprise, revealed himself to be one of the greatest soldiers of history, and like all good generals he had a sound appreciation of what could and could not be done with the various component forces under his command. At Blenheim, after being repulsed four times in frontal

Great to take the next step. In 1759 he formed a brigade of horse artillery armed with light 6-pounder guns, with a view to providing a force of artillery which could manœuvre with and keep up with his cavalry. This he found necessary by virtue of his appreciation of the function of cavalry. Frederick's father had, more or less as a hobby, created an enormous and highly disci-

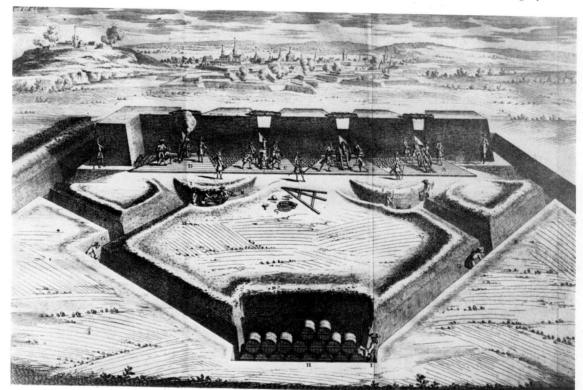

attack, he moved a battery of guns across the River Nebel, and this moving of guns in the course of the battle contributed in no small measure to the day's eventual success.

At the Battle of Malplaquet, which was won at the cost of 12,000 dead, the decisive stroke was again an artillery manœuvre; having penetrated the French centre, Marlborough ordered the 'Grand Battery' of 40 guns to advance into the heart of the French line where, wheeling to face the flanks, they opened a withering fire of case and grape-shot on to the French cavalry who were waiting, behind their infantry, to begin the counter-attack charge. This destruction of the French reserve decided the battle. No doubt had other opportunities offered, 'Corporal John' would have made more use of his guns, but circumstances were sometimes against him; for example at Oudenarde we are told, 'few pieces of artillery were brought up on either side, the rapidity of the movements of both (armies) having outstripped the slow pace at which these ponderous implements of destruction were then conveyed'.

When Marlborough fell from grace after the war, the armies of the world had perforce to wait for another great captain before any further improvement was likely. It fell to Frederick the

plined army which he was too solicitous to hazard in actual warfare. But when the son succeeded his father he found an instrument to hand with which he was able to impress his mark on the whole of Europe. An outstanding soldier and never averse to trying something new, on his accession he found himself in charge of a cavalry force which had been trained to manœuvre into position, then form into line and fire at the halt. While this tactic provided them with excellent firepower, it converted them into little more than mounted infantry, and Frederick, appreciating that movement was the fundamental feature of cavalry action, soon abolished this tactic and trained his cavalry in the use of lance and sword. Having removed their firepower, he had to replace it; he rediscovered Gustavus Adolphus's principles, expanded them and invented horse artillery.

The measure of this innovation can be gauged by the fact that at this time the only mobile artillery in use on the Continent was the 'Battalion Gun', a misguided innovation due to Gustavus which had been perpetuated by those who knew no better. These were light guns dragged along by the marching infantry; they were a species which propounded a dilemma. Either they were light enough not to impede the infantry's rate

of march, in which case they were too light to have much effect when fired; or they were heavy enough to provide a worthwhile lethal effect, in which case they encumbered the infantry and slowed their advance. Usually the bias was to the latter case; had Gustavus lived he would undoubtedly, in due course, have recognized the defects and abolished the battalion gun, but in the event they remained to encumber armies until Frederick's horse artillery showed how mobility and firepower could be brought together.

Frederick's ideas took time to implement, and in the interim the tactics of the day had their effect on the artillery. Frederick's ideas on tactics were easier for the average soldier to assimilate than his ideas on reorganization, and his tactical thinking came to dominate the armies of Europe; drill and

discipline his armies had, and won wars. Drill and discipline therefore became the be-all and end-all of military thinking, and war developed into a matter of position and manœuvre, for with drilled and disciplined troops some elegant manœuvres could now be performed. The defending army selected its position, made its dispositions, and sat there waiting attack. Their artillery was entrenched with it, and it was rarely called upon to move in the course of a battle. The attackers, for their part, secure in the knowledge that nothing short of divine intervention would tempt the defenders from their position, could move at leisure. 'They marched and countermarched, broke into column and wheeled into line with a gravity and solemnity that in our times would provoke a smile', a Victorian analyst wrote. This sort of

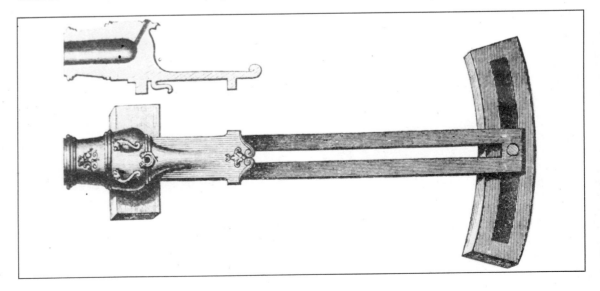

A French 12-pounder
Horse Artillery piece dating
from the Napoleonic Wars.

A British light 6-pounder,
of the type specially
prepared for service in
Canada with General Wolfe.
The axle was widened and
fitted with ammunition
boxes.

35

A French 12-pounder,
contemporary of the
6-pounder, showing the
method of carrying such
vital stores as the hand-
spikes, rammer and sponge,
wash bucket and drag-rope.

A typical naval mortar, as
used on bomb-ketches for
short-range bombardment.

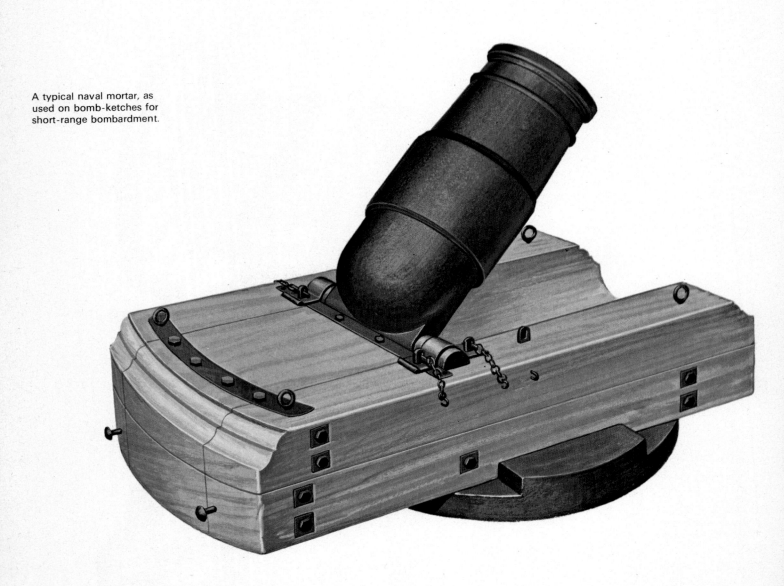

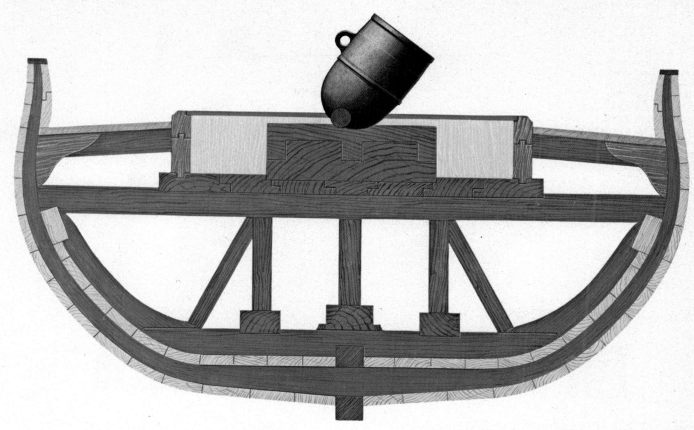

A section through a shallow-draught naval bomb-vessel, with the mortar arranged to fire on the beam. This reveals the heavy reinforcement necessary to withstand the deck blow on firing.

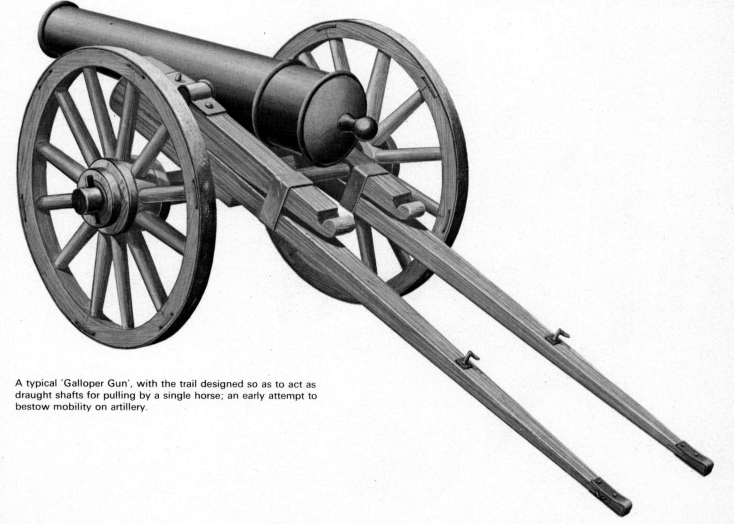

A typical 'Galloper Gun', with the trail designed so as to act as draught shafts for pulling by a single horse; an early attempt to bestow mobility on artillery.

The complicated arrange-
ment of tackle necessary to
control and secure a naval
gun of the eighteenth
century. The light 'side-
tackle' was for securing the
gun when not in action;
the heavier 'breeching rope'
was the recoil check.

An early pattern of naval
truck carriage, showing the
method of construction.
The basic structure was of
elm, chosen for its ability
to absorb shock and its
resistance to splintering
when struck by enemy shot.
The whole carriage is
rigidly held by through-
bolts, while the wheels are
in two halves, with the
wood grain opposed.

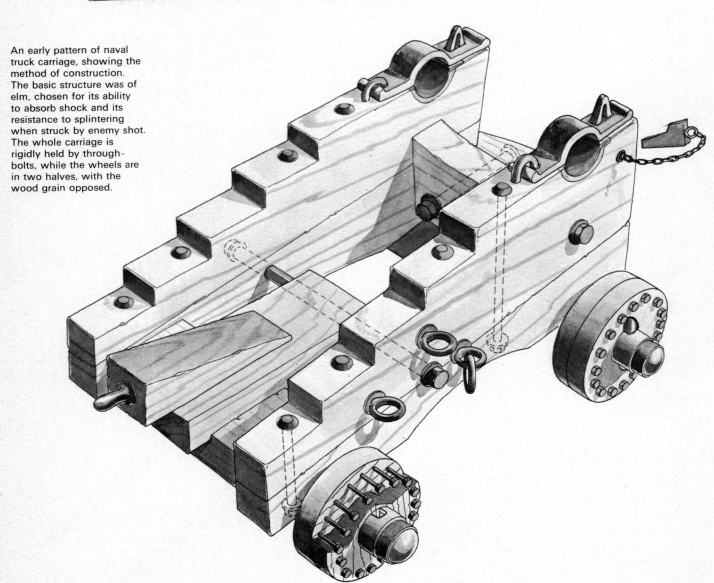

armed gavotte reached its zenith at Fontenoy with Lord Charles Hay's infamous invitation to the French to fire first. But the system was accepted as the only method of fighting, and it remained the doctrine until Napoleon reintroduced mobility, which upset several people. 'In my youth', complained an elderly Prussian officer, 'we used to march and countermarch all summer without gaining or losing a square league, and then we went into winter quarters. But now comes an ignorant hot-headed young man who flies from Boulogne to Ulm, and from Ulm to the middle of Moravia, and fights battles in December. The whole system of .his tactics is monstrously incorrect.'

The general result of this dilatory tactical system was to produce a tendency to improve the accuracy and effect of artillery fire to the detriment of mobility, leading to the gradual adoption of heavier guns of larger calibre. But in spite of this trend there were one or two attempts to produce more practical weapons from time to time, attempts which prevented artillery from sinking entirely from sight. One rather eccentric innovator was the Chevalier Folard who decided to design

a lightweight gun and produced a short 24-pounder. With a 28-inch barrel it weighed only 15 cwt, a startling change from the conventional 24-pounder of the day which was 11 feet long and weighed 45 cwt. Unfortunately when constructed and fired, it blew up; this regrettable result so upset the good Chevalier that he came to the conclusion that artillery was incapable of any improvement, and he proposed the complete abolition of the arm, replacing it with mobile ballista and catapults.

Perhaps nothing better illustrates the poor state of artillery at this time (1723) than the fact that Folard's ridiculous proposals were seriously considered. Even such an astute intelligence as Benjamin Franklin was swayed by Folard's arguments and in later years urged upon General Lee the suppression of artillery and the reintroduction of archery.

However, this was the lunatic fringe. At the same time as the Chevalier was advocating a return to catapults, others, more versed in artillery fundamentals, were also taking a look at the lightweight gun. The first move was in Germany in about 1725 when a number of 8-pounder and

Folard's disillusionment with the state of artillery led him to advocate equipping the troops with this *catapulte de campagne* instead.

(*Right*) Marshal Saxe suggested provision of this 'Amusette' in considerable numbers, but the idea failed to catch on.

(*Below*) Another idea which failed was M de Bonneville's mobile 1-pounder breech-loader.

(*Bottom*) Cornelius Redlichkeit's disappearing gun carriage; on recoiling, the small carriage runs down to inclined plane, counterbalanced by the heavy roller. From Scheel's *Memoirs d'Artillery* published in Denmark, 1777. (*Royal Artillery Institution*)

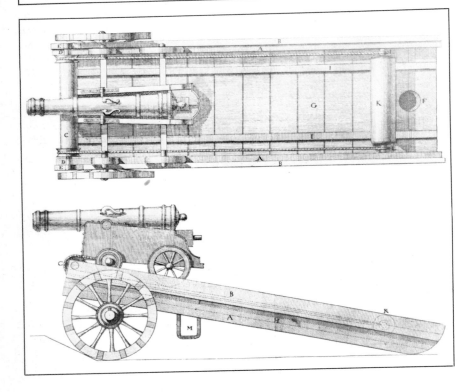

4-pounder guns were mounted so that they could be brought rapidly into action and fired without detaching them from the horse. Their firepower was inferior but the balance of advantages was in their favour, lightness compensating for poor lethality. What the horses thought about the idea is not on record. The 'Galloper Guns' which appeared in the 1740s were a further and more practical development of this idea; the carriage was made with shafts which could act as a trail when the gun was in action.

Unfortunately, while the galloper gun calls up a dashing image the reality was less stirring. The design well illustrates the confusion between lightness and mobility. The gun was light and mobile, no doubt of that. But the flaw in the system was that while the guns were capable of rapid movement they did so at some disadvantage; the ammunition was on heavy carts and the gunners were mostly on foot. So for all the lightness, mobility was still absent.

Marshal Saxe was the next to try his hand; he had a high opinion of the power of artillery but a poor one of its mobility. 'It is unlikely that the artillery will ever move faster; it is impossible that it will ever move slower,' he is reputed to have said. And to remedy the deficiency he proposed the 'Amusette', a species of heavy musket firing a half-pound ball and drawn by hand, to be distributed in large numbers across the front of the battle. Nothing seems to have come of this suggestion, but it was echoed a few years later (1762) by another Frenchman, M. de Bonneville. He proposed a 1-pounder breechloader which, according to him, could be loaded and fired on the move. This idea also never seems to have reached the field of battle.

In these years of tactical ferment, one is entitled to ask if there had been any technical advance in the material of artillery. Fortunately, here the picture is brighter. This side of the matter was in the hands of the gunners themselves, and, with a certain faith in the rightness of their calling they applied themselves to improving the tools of their

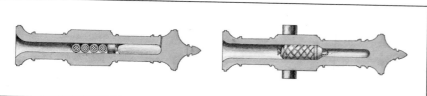

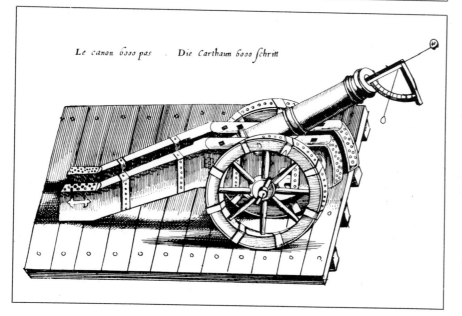

Le canon 6000 pas   Die Carthaun 6000 schritt

(*Top and bottom*) Seventeenth-century drawings showing the use of the gunner's quadrant.

(*Centre*) Two Russian 'Schuwalon' cannons of the early eighteenth century.

no way round that problem, but there were other fields to be explored.

The gun carriage, two wheels joined by an axletree and with a trail to support the weight and the shock of firing, had superseded the gun cart in the fifteenth century, and in about 1500 came the first gunnery instrument–the gunner's quadrant. This is reputed to have been invented by the Emperor Maximilian I, and was no more than a 90-degree quadrant with one side extended, carrying a plumb-bob. Since degrees were not yet known, the quadrant was arbitrarily marked off in 'points'. By placing the extended side in the cannon's bore the weapon could then be elevated or depressed until the plumb-bob indicated the desired point to achieve the required range. With the gun horizontal the plumb-bob reached the end of the scale, from whence comes the term 'point-blank'.

Having a scale of points and equating them to ranges demanded the production of some form of table of ranges and elevations, and this was something the gunner had to find out for himself, for guns were individual weapons and not mass produced. All sorts of minor variations in dimensions could be found between two nominally identical guns, and in addition every gunner was idiosyncratic about how much powder he used, how he rammed it, whether he used a wad and so forth. Thus it was necessary for him to take his gun out and actually fire it at the various points on the quadrant, measuring the result of each shot and recording it for his future use.

The actual task of elevating the gun was done by heaving it up or down by the use of levers or handspikes, inserting wooden blocks beneath the breech to hold it at the required angle; the blocks were soon refined into a wedge which gave more precise control, and the ultimate system came in about 1578 when John Skinner, 'one of the Queen's Majesty's Men' invented the elevating screw, which gave finer control. Some early guns, as can be seen from the illustrations, used an arc perforated with holes to position the breech end of the gun, but this was only suited to the lighter types of weapon. Whichever system was used, there were, as yet, no sighting arrangements; the gunner merely looked over the line of the gun, elevated by means of quadrant and range table, and hoped for the best.

In the ammunition field, Stefan Batory, King of Poland, is credited with the introduction of red-hot shot in 1579. This device, more useful against ships and property than against men, required some dexterity on the part of the gunners to fire it without doing themselves harm. The iron shot was heated to redness in a furnace; the gun was loaded with a charge of powder and a tight-fitting dry wad rammed down on top; then, with great rapidity, a wet wad was rammed down, followed by the red-hot shot, whereupon the gun was touched off–before the shot burned its way through the wads and did the job itself. Primitive as it sounds, it remained a standard item of ammunition until the smoothbore gun disappeared

trade. No matter that the generals and marshals were incapable of handling the guns or appreciating their worth; when the day came that their talents were recognized, the gunners would not be found wanting. The guns themselves were long and ponderous still, due to the powder. Slow burning, it demanded a long and thus heavy barrel to develop its full force. There seemed to be

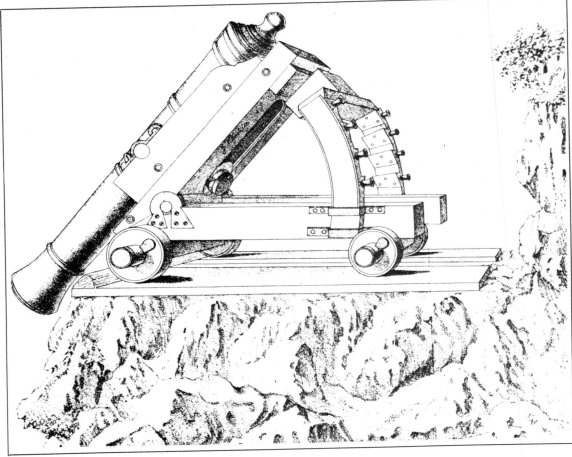

Lieutenant Kohler's 'Depressing Carriage', developed during the Siege of Gibraltar to allow the guns of the fortress to shoot downwards.

An early breech-loading naval gun of the type used in the early sixteenth century, mounted on a simple bed, formed at the rear end to support the removable chamber section.

from the scene in the nineteenth century.

In 1588 comes the first record of the use of hollow cannon balls filled with gunpowder, these being used to shell Bergen-op-Zoom, thus trans-lating the explosive effect of the powder to the target and bringing new meaning to Bacon's ob-servation that 'These substances can be used at any distance we please, so that the operators escape

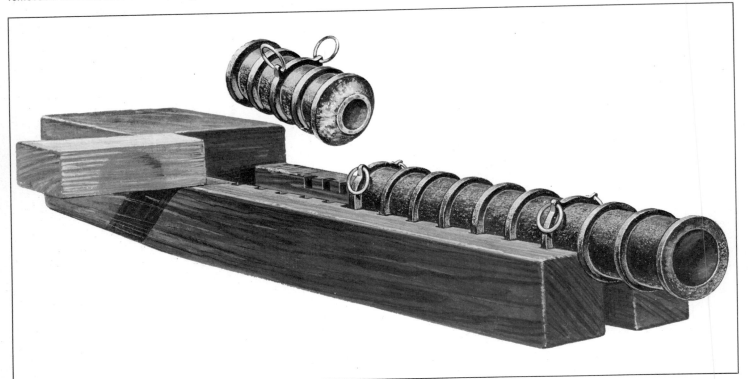

(*Above*) A Spanish gun of 1628, ornately decorated; one of the huge collection in the Military Museum of Lisbon.

(*Right*) British muzzle-loading 32-pounders on garrison standing carriages, mounted on the ramparts of Plymouth Citadel.

and then burn away to ignite the shell contents at the end of the trajectory, but his idea was not followed up for many years; one drawback to the development of such a 'time fuze' was the simple question of calibrating such a device when no accurate method of measuring small intervals of time existed.

At sea the use of ordnance had made a slow start. Sea battles for the most part were simple and bloody affairs in which one ship grappled to another and the crews fought it out hand to hand, and the use of cannon was confined to short-range fire with peterara and the like, loaded with 'langridge' – scrap metal and small stones – to repel the boarders. It was not until the middle of the fifteenth century that the use of cannon as offensive arms, to reach across the intervening water and damage the enemy before he could come to grips, became a standard practice. Among other reasons, the bulk and weight of the contemporary long-ranging gun was a considerable problem, and not until the general introduction of cast-iron guns and corned powder allowed the development of handier weapons did the sailors take kindly to burdening their craft with cannon.

By the time of Elizabeth I the seagoing cannon was an accepted item, and so far as the gun itself was concerned its advance paralleled that of land artillery. The principal difference lay in the question of adapting the weapon to the ship – the gun carriage or mounting. The first ship-board guns appear to have been simply barrels laid in a wooden trough, the trough being fixed to the ship and the barrel free to recoil in it, controlled to some degree by ropes or chains. This was later changed, when it was appreciated that increasing the mass of the recoiling parts decreased the violence of recoil, to firmly attaching the cannon to the trough and allowing both to recoil. Then, some time in the sixteenth century, came the addition of wheels, or trucks, to the trough, and from this rough beginning the 'truck carriage' or 'ship carriage' evolved.

The truck carriage was, in fact, far from the perfect answer, and even its champions had to admit that it had its defects. The system of controlling recoil by the 'breeching rope' was primitive; if the tackle securing the gun broke loose in a storm, the task of catching and securing the runaway was extremely hazardous and if not done quickly could well lead to greater disasters. More than one ship lost with all hands had her foundering attributed to the guns breaking loose in a storm. The attachment of the breeching rope and running-out tackle invariably caused the gun to jump on firing, to the detriment of accuracy, and the sailors had to step lively to avoid being struck by the recoiling gun or caught up in the festoon of ropes and tackle. But having said all that, it had to be admitted that the truck carriage was simple, robust, easily repairable by the ship's carpenter and did its job. Since nothing better offered, the truck carriage was to stay in service until the nineteenth century with very little improvement.

all hurt from them, while those against whom they are employed are suddenly filled with confusion.' The operators did not entirely escape all harm though; the explosion of the powder at the target was brought about by internal friction when the shell struck its target, and often an equal friction was developed when the shot was launched, so that the explosion took place at the beginning of the trajectory instead of at the end. One Sebastian Halle proposed a way round this in 1596 by the use of a wooden peg inserted into the shell and containing a filling of gunpowder, which would be ignited by the explosion of the charge

CHAPTER 3

Scientist to Engineer

The year 1742 was a notable one in the development of artillery, since it saw the first major step towards understanding the science of gunnery and ballistics. Benjamin Robins read his paper 'New Principles of Gunnery' before the Royal Society.

Robins was a brilliant mathematician and incessant experimenter, and one of his greatest achievements was the invention of the ballistic pendulum. This device, rough as it was, was the forerunner of all modern ballistic measuring devices and, it has been said, was as much a scientific milestone as Galileo's telescope or Watt's steam engine. The calculations done and formulae derived by Robins from his experiments with the pendulum were used for over 100 years, until the perfection of electrical systems which allowed more precise measurement.

Before Robins's day the measurement of a gun's performance was entirely empiric. The gun was fired, the ball went so far; with a different charge it went a greater or smaller distance. Beyond that nothing could be said. Robins saw that if the ball could be made to transfer its energy to something capable of being more easily measured,

he would be able to calculate the energy and velocity of the projectile and thus have some definite measure of the gun's performance. He constructed a framework from which was suspended a heavy pendulum, the arm of which was an iron rod carrying a massive baulk of timber. Attached to the bottom of the pendulum was a measuring tape running past a marker. The gun to be tested was set up in front of the pendulum and the ball fired so as to strike the wooden baulk. On striking, the ball caused the pendulum to swing, pulling the measuring tape past the marker. Since the weight and length of the pendulum had been carefully determined, knowledge of the distance it had moved could now be used to calculate the striking energy and velocity of the ball. By performing the experiment at various distances, Robins was able to determine the loss of velocity as the range increased, and from this he derived a formula for the motion of the projectile which took into account both the force of gravity and the effect of air resistance.

Robins was, indeed, so far ahead in his appreciations of gunnery problems that had sufficient attention been paid to his suggestions and practical applications derived from them, the technical development of ordnance would have moved ahead at a much faster rate, and the technical revolution of the 1850s would have had a much sounder

A howitzer of the late eighteenth century, contemporary with that shown on page 55, showing the method of construction.

(*Above*) Some early projectiles: from left to right, a 6-pounder 'spherical case' fitted with a 'wood bottom' to ensure that the fuze was in the right place after loading; a common shell; and a grape-shot.

(*Right*) A highly ornamented 'Fish Gun' taken from the Palace of the King of Oudh on his deposal in 1856. (*Royal Artillery Institution*)

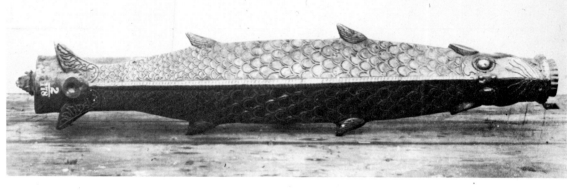

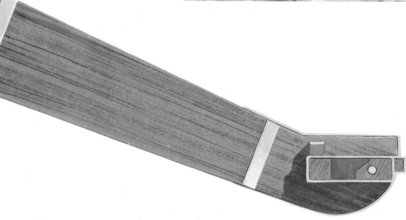

foundation from which to spring. As an example of his forward thinking, his observations on rifled ordnance are worth repeating:

'Whatever state shall thoroughly comprehend the nature and advantages of rifled barrel pieces and, having facilitated and completed their construction, shall introduce into their armies their general use, with a dexterity in the management of them; they will by this means acquire a superiority which will almost equal anything that has been done at any time by the particular excellence of any one kind of arms; and will perhaps fall but little short of the wonderful effects which histories relate to

have been formerly produced by the first inventors of firearms.'

The ballistic pendulum and Robins's papers spurred a number of experimenters into life, notably in the direction of trying to measure the velocity of the projectile. In 1764 a Lieutenant de Butet of the Italian service produced a machine which consisted of a wheel revolving at a mea-sured rate. When the gun under test was fired the shot broke a string which allowed a marker to contact the rim of the wheel, the revolution of which was stopped by the shot striking a butt a measured distance away. Thus the mark on the wheel's rim represented the time of the shot's flight over the measured distance; since the peripheral speed of the wheel was known, the velocity of the shot could be determined. Needless to say

This ornate bronze
9-pounder is fitted with
axle-tree seats for two of
the gunners.

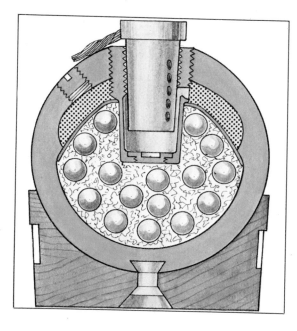

Boxer's improved shrapnel
shell; Shrapnel's original
design mixed musket balls
and powder, a combination
liable to accidental
explosion. Boxer separated
the two by a thin diaphragm,
improving both safety and
functioning.

the machine, while ingenious, was hardly of the
degree of precision needed to deal with the
velocities in question, and the results it gave were
of doubtful accuracy.

Stemming from this came a number of
'machines of rotation', all of which were based on
similar principles. Matthei's machine of 1767 is a
good example of the class. Here the revolving
wheel carries a circumferential screen of paper.
Were the wheel stationary a ball fired across it
would produce two holes in the paper 180 degrees
apart. With the wheel in motion at a measured
rate, the angle between the two holes would give
a measure of the wheel's movement while the ball
was crossing it, thus allowing velocity to be calculated. While the theory was sound enough, the
dimension of the wheel, the difficulty of obtaining
an absolutely regular speed, and the absence of an
accurate system of measuring small periods of
time all added up to some degree of error.

This activity in the laboratory was paralleled by

activity in the field. Frederick's horse artillery had shown its paces and worth and, slowly, the rest of Europe began to appreciate the value of mobile artillery. The only problem was that of how best to go about organizing it. The greatest difficulty was making sure that guns, gunners and ammunition were so inseparably welded together that they all arrived in action at the same place and the same time. There were five systems of managing this open to selection. Firstly the *detachment system* where all the gunners were on horseback; then the *off-horse system* in which the gunners were mounted, some on the off-horses of the gun teams and some on the off-horses of the waggon teams; then the *limber system* in which seats were provided on the gun carriage and limber; fourthly, the *car system* in which the gunners were provided with a special carriage for their transport; and finally the *waggon system* in which they all rode on the ammu-

marked, it was accepted that field artillery could have somewhat heavier guns since they would not be expected to move at the gallop.

These factors, and many others, occupied many minds, and the Austrian Army was the first to produce a mobile artillery force. First organized on the car system, in 1778 they were reorganized into the limber system, with the gunners riding on gun carriage and limber seats. In 1776 the French artillery was radically overhauled by Gribeauval, one of the greatest artillery experts and reformers of history, but due to political pressures he was unable to push all his proposed reforms through, and the French artillery, while efficient in action, got there slowly since its gunners were still afoot. It was not until 1792, under the pressure of war and the urgings of Lafayette that horse artillery and limbers were adopted.

It was this same rumble of war which crossed

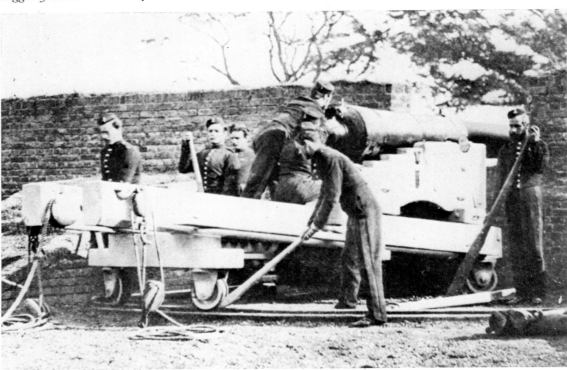

Traversing an eighteenth-century coast artillery piece by means of handspikes.

nition waggons. The provision of ammunition for the immediate service of the gun was solved by the invention of the limber shortly after the beginning of the Seven Years War; this took the form of a light wheeled box carrying a small supply of ready-use rounds, to which the gun was hooked, the whole equipage then being drawn by the horse team.

The size of the horse team, of course, decided the size of the gun; indeed, from the middle of the seventeenth century to the beginning of the twentieth, the first criterion in field gun design was weight–could it be moved by the standard horse team? Six horses, two abreast, was accepted as being a convenient size of team, and six horses can gallop with 30 cwt, which put an upper limit on any horse artillery gun. When the distinction between horse and field artillery became more

the Channel and aroused the English. In 1788 the Duke of Richmond, then Master-General of the Ordnance, gave instructions to prepare and equip a number of guns 'capable of accompanying cavalry in the field'. After much debate and the formation of a committee, three schemes were placed before the Duke, and in 1793 he authorized the formation of two troops organized on the detachment system, with the gunners on horseback.

With this general reorganization, artillery was at last in a position to make itself felt on any point of the battlefield, and in the Napoleonic Wars which followed, their handling was perfected and their firepower tested in numerous engagements. Concurrently with the improvement in the guns, there had been an equally important step taken in ammunition. In 1787 a Lieutenant Henry Shrapnel

A 13-inch mortar being brought into action at drill.

A siege gun stands in its revetment, buttressed by barrels of earth. On the right are the rammers and cleaning sponge.

was stationed at Gibraltar, recently under siege by the Spaniards. Small gunboats, attempting to come close and bombard the defenders, were undaunted by the firing of ordinary shot. A Captain Mercier of the 39th Regiment conceived the idea of firing 5·5-inch mortar shells from a 24-pounder gun, fitting the bomb with a short-burning fuze so that it burst in the air, spreading its fragments about and making life much more hazardous for the boats. Shrapnel made a number of trials and finally devised a new method of defence by filling an 8-inch mortar shell with 200 musket balls and a small charge of powder. He then fitted a powder fuze to the shell, and on 21 December 1787 demonstrated, before General O'Hara, the commander of the fortress, that with the fuze accurately cut he could burst open the shell some few feet above the surface of the water, whereupon the musket balls were released to devastate a 30-foot circle. When Shrapnel left Gibraltar in 1790 he continued working on his idea, finally making a formal suggestion to the Master-General of the Ordnance in 1792. A committee was formed to examine his proposals, which, after nine years, reported that 'the effect

appears to be very considerable' but 'do not take upon themselves to decide upon the policy of introducing it for general service . . .' However the Board of Ordnance were quite ready to assume this responsibility, and in August 1803 Shrapnel was ordered to the Carron Foundry in Falkirk to supervise the production of the first of his 'Spherical Case Shot'. By the end of the year no less than 74,442 shells had been manufactured, and the projectile was launched on a long and lethal career.

The first use of this new weapon appears to have been in a small action at the Batavian settlement of Surinam on 30 April 1804, the artillery commander reporting that they 'had so excellent an effect as to cause the Garrison of Fort Amsterdam to surrender at discretion after receiving the second shell. The enemy were so astonished at these shells as not to be able to account how they apparently suffered from musketry at so great a distance as 2,050 yards.' This was, indeed, the significant feature of Shrapnel's invention–that he could virtually duplicate musketry fire at any range within reach of cannon–and since there have been so many misconceptions of his principle, it is worth examining the idea more closely.

Case or canister shot had been known since the fifteenth century. This was no more than a thin metal canister–the 'case'–filled with musket balls and shot from the cannon so that the explosion of the cartridge split open the case and launched the balls from the muzzle in the manner of a shotgun charge. As a close-range anti-personnel weapon it is without equal, but at ranges much more than 300 yards it is useless. Grape-shot–heavier lead balls in a canvas bag, tied so as to resemble a bunch of grapes–operated in similar fashion, but due to the heavier balls it took effect to a longer range. Shrapnel's idea was to carry the shell to the enemy and then burst it so as to have the same effect as case but at extreme ranges. The significant point is that the shell contained only sufficient gunpowder to open it and release the balls, which were endowed with the forward velocity of the shell alone. There was no question of using the force of the powder explosion to drive the balls at any greater velocity; furthermore the balls would continue on the same trajectory as the unexploded shell would have done, so that predicting their point of impact was relatively easy. It is this use of minimal force to open the shell and the reliance on the momentum imparted by the shell's flight which is the feature distinguishing Shrapnel's idea from other systems of loading musket balls into shells which have, from time to time, been advanced as precursors of his idea.

In spite of his title of 'Spherical Case Shot' the shells inevitably came to be called by the inventor's name, and finally, in 1852, his family made application to the War Office to have 'Spherical Case Shot' officially named 'Shrapnel Shell', 'from

Fuentes de Oñoro; during a tactical withdrawal of Craufurd's Light Division, Ramsay, with two guns of Bull's Troop, was halting periodically to give covering fire and keep the pursuing French cavalry at a respectful distance. At one halt, lingering too long, he was cut off just as the guns had been limbered up, by a host of chasseurs who rode in from a flank. The gunners drew their swords, spurred their teams to a gallop, and with Ramsay at their head charged and cut their way into the French cavalry, eventually aided by two squadrons of dragoons who turned back on seeing 'their' artillery in trouble.

A technical innovation which might be instanced was the use of supporting fire for infantry at San Sebastian in 1813. For the first time the guns continued to fire over the heads of the advancing infantry, only lifting their fire from the target when the final assault was to be made, 'a

This bronze gun, captured in the Crimea, stands on an unrepresentative type of carriage; the cascable is missing, having been removed to provide the bronze from which Victoria Crosses are minted.

the circumstances that other nations have long since done this honour of invariably attaching his name to the weapon and because the family have not the means to afford the expense of erecting a monument . . . but which such a distinction would be the means of representing'. As a result an army order was published on 11 June 1852, directing that the projectile be henceforth known as 'Shrapnel Shell' in honour of its inventor.

The combination of shrapnel shell and horse artillery brought a new dimension to warfare and a new importance to artillery which was increasingly exploited during the Napoleonic Wars. As good fortune would have it, the times also brought forward a number of brilliant and resourceful young officers to command the new artillery and many feats, both of bravery and of technical innovation, were seen. An example of the former was the exploit of Captain Norman Ramsay at

feat which proved the skill of the gunners as well as the accuracy of their pieces', and which can be justly claimed as the ancestor of the barrage fire which came into prominence just over 100 years later.

The action in the Peninsula brought a good deal of operations in the mountains of Spain and Portugal. The horse artillery had showed its worth in being able to keep up with troop movement, and the infantry operating in the peaks and pine trees demanded some sort of firepower which could accompany their vertiginous scrambling. As a result, a number of 3-pounder guns were built, capable of being dismantled, carried piecemeal on mule-back, and then readily reassembled for firing.

The introduction of this 'Mountain Artillery' meant that there were now four distinct types of artillery with the field army: mountain, horse, field and, inevitably bringing up the rear due to its weight, siege. There was also, of course, the static 'Garrison Artillery' emplaced in forts for the defence of frontiers and coastlines. We have said little of siege artillery, since during the years it had seen little technical advance. Since it was the role of the more mobile branches to support during open battles, the siege guns were allotted certain specific responsibilities which rarely entered play until a formal siege developed. Once this event took place, the 'Siege Train' was moved forward and emplaced, and their task thenceforward was defined as keeping down the fire of the besieged and protecting the besiegers; ruining the defences and preventing their repair; destroying the stores and magazines within the besieged place; and the creation of breaches in the defences through which the assaulting troops could gain access.

During the seventeenth and eighteenth centuries fortification, from being a haphazard affair of ramparts and ditches, had blossomed forth into one of the foremost of military sciences, replete with its own vocabulary and mystique. But at the same time the art of besieging had been reduced almost to a formula and indeed, Cormontaigne, one of the major prophets of fortification, actually drew up a table showing how long a work could stand siege, depending upon who had been the architect responsible for its construction. Thus Vauban's first system might be expected to hold out 19 days, his third system for 26 days, Cohorn's system 21 days and so forth. The besiegers arranged themselves outside the fort, brought their guns into play, and began the formal moves of sap trench and parallel; and provided everybody played by the rules, Cormontaigne wasn't far wrong. But if anyone circumvented the rules, then

(*Right*) An eighteenth-century Maltese gun. (*Royal Artillery Institution*)

(*Below*) Some projectiles which illustrate early attempts at stabilizing the projectile: from left to right, a Whitworth shell for use with a hexagonal bore; a Sawyer shell with ribs to engage in the rifling; and a breech-loading shell using a brass bottom plate to engage in the rifling. This latter was intended as an armour-piercing shell and has a concave head formed into cutting edges.

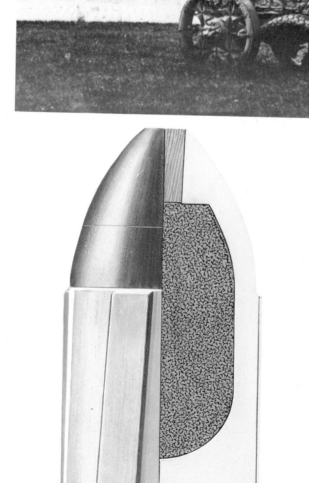

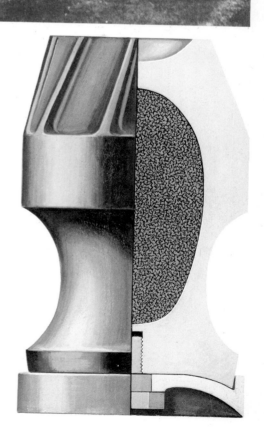

A Lancaster oval-bore gun, one of many designs submitted in the 1850s when various systems of rifling were being investigated. (*Royal Artillery Institution*)

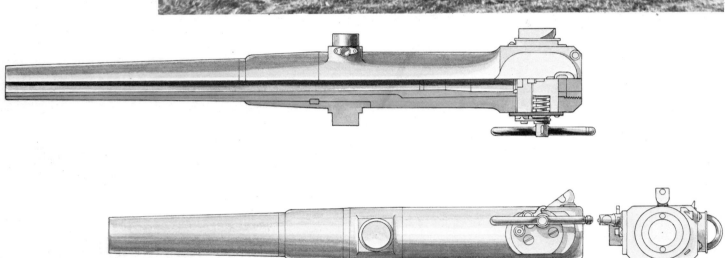

A Russian 'Obuchov' 24-pounder with sliding block breech, based on the Krupp design.

anything might happen. Ulm, in 1706, was breached by stealth, by a party of officers disguised as young women, an activity which made nonsense of tables based on the geometrical calculations of the fortress engineer.

As a result, the siege artillery became more ponderous in order to propel projectiles capable of doing the necessary damage. Mortars were highly regarded in this role, since they could develop 'curved fire' to drop explosive shells within the defences, while guns were used to fire directly at the fortifications in order to make the desired breach. As a result of numerous sieges of Napoleonic times more interest began to be taken in siege gunnery and the more-or-less empiric shooting of former days was taken under review. Piobert, a French experimenter, conducted numerous trials at Metz in 1834 against obsolete fortifications, renewed his tests in 1844, and was associated with more trials at Bapaume in 1847. At Bapaume a number of obsolete works were to

be demolished to make way for new, and the opportunity was taken of testing the fire of artillery. From these trials was developed a scientific system of battering a wall with solid shot to break a section up and then firing explosive shell in order to get rid of the fragments which were by then smothering the effect of the shot; this technique became known in later years as the 'pick and shovel' method, the shot acting as the pick and the shell as the shovel.

In the long peace which overtook Europe after the defeat of Napoleon, except for a few isolated experiments such as those of Piobert, little of note occurred and the world's armies slumbered. Then in 1854 came the Crimean War, a small affair in itself but a war which was the catalyst provoking a technical upheaval which has lasted ever since. With one or two exceptions—Inkerman, for example—the actions of the Crimean War were largely sieges; the Russians besieging the Turks on the Danube and the Allies besieging the

The ammunition limber for the howitzer below; its size was governed by the weight capable of being drawn by a two-horse team.

A light howitzer of 1780. Although a relatively small gun, the weight of shot developed a heavy recoil, necessitating a strong carriage.

The 20-pounder Parrot
rifle of 1861. Of 3·67-inch
calibre it had a range of
1,900 yards and was
notably accurate. Small
numbers were used during
the American Civil War.

An Armstrong-Whitworth
cannon, used in the
American Civil War.
Designed by Whitworth to
use his hexagonal bore, and
built by Armstrong on his
'built-up' system, the breech
mechanism is an unusual
type in which the breech
screw locked on to the
outside of the gun breech.

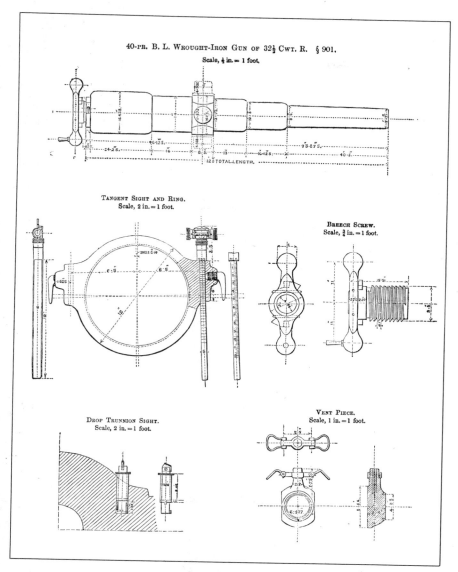

40-PR. B. L. WROUGHT-IRON GUN OF 32½ CWT. R. § 901.

Scale, ¼ in. = 1 foot.

TANGENT SIGHT AND RING.
Scale, 2 in. = 1 foot.

BREECH SCREW.
Scale, ¾ in. = 1 foot.

DROP TRUNNION SIGHT.
Scale, 2 in. = 1 foot.

VENT PIECE.
Scale, 1 in. = 1 foot.

The official approval of the Armstrong 40-pounder RBL gun included this drawing which gave details of the gun sights and breech vent-piece.

Armstrong's rifled breech-loading 110-pounder of 7-inch calibre, 1861. (*Royal Artillery Institution*)

Russians at Sebastopol, and hence artillery took a prominent part. At the same time the Crimean War introduced the war correspondent, and it was fortunate that the first of the species was a journalist of sufficient integrity to report freely on what he saw, so that the newspaper readers of the world could assess the performance of the Allies and their weapons in a manner hitherto unknown.

Now the Victorian Age was an age of invention; anyone who doubts this should visit his national Patent Office and examine their files from 1850 to 1899. This was due to a combination of an inquiring spirit and the impetus of the Industrial Revolution. Outside the field of electronics, most of the machinery we know today had its beginnings in a sketch drawn in the latter half of the nineteenth century; frequently the sketch got no further, since inventiveness outstripped mechanical and metallurgical ability and knowledge, but the seeds were there. The sudden national enthusiasm for the Crimean War in England, coupled with the reports of inefficiency and antiquated equipment, set pencils scudding across drawing-boards up and down the land, and from it all came a sudden interest in ordnance engineering which engaged some of the best mechanical brains in Europe. For Krupp, Erhardt, Schneider, Armstrong, Whitworth, Lancaster and many others, the stimulus of the Crimean War was to lead them into new paths and make their names known throughout the world.

It should not be supposed (though it often is) that these gentlemen had to fight against the iron hand of reaction in presenting their ideas to the soldiers; far from it. The soldiers knew well enough about the deficiencies in their equipment, but derisory funds and lack of interest by good engineers had denied them any useful progress. The British Army were well acquainted with the theoretical advantages of rifled ordnance and Robins's dictum was well appreciated; they had, previous to the outbreak of war, let it be known that they would look favourably on any suitable design put forward. But the only people who responded were inventors whose agility of mind far outstripped their technical competence, and what few ideas came forward were unacceptable. The advent of rifled small arms, with their increased range and accuracy, made it imperative that artillery should also reap the advantages so that it would not be outranged by the enemy infantry before it could even get the guns into action.

Numerous systems had been tried during the early years of the century. In 1821 a Lieutenant Croly of the 1st Regiment of the British Army had proposed a rifled breech-loading gun using a lead-coated projectile. A similar idea, rather better thought out and engineered, was developed in Sweden by Baron Wahrendorff in 1846, while in 1842 a Colonel Treuille de Beaulieu of the French Army put forward a system of rifled muzzle-loading and in 1845 a Major Cavalli of the Sardinian Army proposed a two-grooved rifled barrel with a ribbed shell to suit it. While some of these ideas were built and tried, none appeared to be sufficiently serviceable to warrant the enormous expense of equipping an entire army. But with the Crimean War under way, more designs appeared, money became available, and eventually, in 1854 the British Government took the decision to order some cast-iron guns to be made on the

Lancaster principle and shipped off to Sebastopol to be tried in action.

The Lancaster system could hardly be called rifling in the generally accepted sense of the word; the gun was still, in fact, smoothbored, but instead of the bore being the usual cylinder it was oval in section and twisted so that it made one complete turn in 30 feet. The projectile was elliptical in section and planed on the skew to match the twist of the bore.

The guns were not a success. The shot had a tendency to jam in the bore and damage the interior surface, and the accuracy, poor to start with, deteriorated as a result. They were soon withdrawn from service. In the same year Mr William Armstrong had appeared on the scene with some radically new ideas, proposing a completely new type of weapon, and this was soon put to the test.

Armstrong's gun was designed from first

(*Above*) Engstrom's gun carriage, yet another early attempt to control recoil. (*Royal Artillery Institution*)

(*Right*) An early Krupp steel breech-loader, captured from the Boers in South Africa.

Breech of the Krupp gun showing the early form of operation by a slow screw.

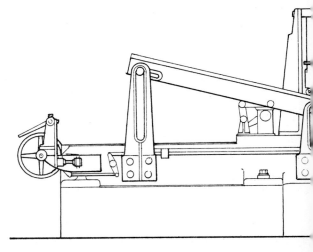

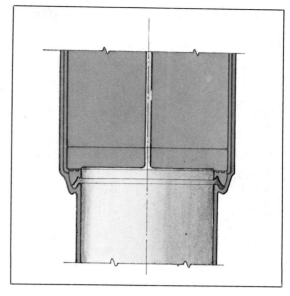

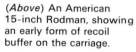

(*Above*) An American 15-inch Rodman, showing an early form of recoil buffer on the carriage.

(*Above right*) The Broadwell Ring obturation system was used extensively with the Krupp Sliding Block system when used with bagged charges. It was later abandoned in favour of using brass cartridge cases, but was revived, in a much improved form, during the Second World War.

(*Below*) A rifling machine; the cutting head is attached to a shaft, inserted into the gun bore, and withdrawn, during which withdrawal stroke the cut is made. As the bar is withdrawn, by moving a supporting carriage across a firm base, a roller on the cutting shaft moves on a 'copying bar' and imparts the desired twist to the cutter.

principles, instead of taking an existing smoothbore and modifying it, which was the usual practice. To begin with he abandoned cast iron as being an antediluvian material and proposed manufacturing his guns from wrought iron. He also proposed building up the gun by shrinking a succession of tubes–or 'hoops'–around the basic barrel, so as to build up the requisite thickness of metal in a manner proportional to the internal pressure to be expected at each point along the gun. Furthermore the shrinkage of each hoop would place the previous hoop under compression, thus providing the metal with greater resistance to the internal force of the exploding cartridge. The barrel was rifled with a large number of shallow grooves, spiralling at a rate of about one turn in 30 or 40 calibres–the twist varied with different calibres, depending on the weight and velocity of the projectile–and the projectile itself was coated with lead so as to engage with the grooves. (It will be appreciated that this part of his design was based, to some extent, on Croly's and Wahrendorff's early ideas.)

Finally, the gun was to be breech-loaded; the rear of the barrel was closed by a 'vent piece'

dropped in in a similar manner to the old peterara, and clamped in place by a large screw, this having a hole through its axis to facilitate loading.

Armstrong delivered a 3-pounder for trial in October 1855 and the Ordnance Select Committee reported favourably upon it. But it was thought that to adopt the Armstrong system forthwith would be inequitable, and for some years comparative trials were carried out with a number of guns rifled on different systems in order to discover whether any of them showed more promise than Armstrong's design, and, very important, whether equal results could be obtained by using something less expensive–for the Armstrong gun's technical novelty and complexity was reflected in its price. Eventually, on 16 November 1858, the Special Committee assembled to conduct the trials had 'the honour to recommend the immediate introduction of guns rifled on Mr Armstrong's principle' and this recommendation was accepted.

Armstrong deeded all his patents to the Crown in January 1859, and in the following month he was appointed 'Engineer for Rifled Ordnance' to the War Department. In November 1859 he

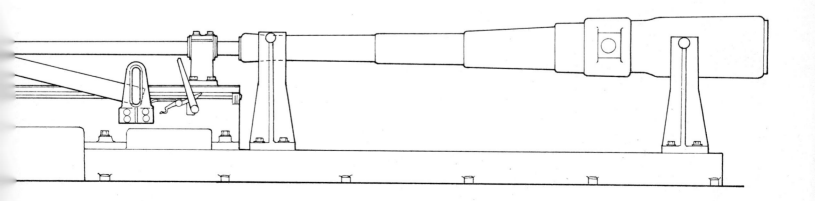

became Superintendent of the Royal Gun Factory at Woolwich so that he could devote all his time and energy to the production of the guns.

All this activity did not go unnoticed elsewhere. In France the Emperor urged his artillery to attend to the rifled gun question and they adopted de Beaulieu's system. In this the barrel of a bronze smoothbore was cut with six deep spiralling grooves, and the projectile had rows of soft metal studs let into its surface. The gun was muzzle-loaded, the shell studs being entered into the grooves at the muzzle and the shell rammed down on top of the charge in the normal way. The explosion of the charge then caused the shell to ride up the grooves and thus pick up rotation on its way out.

The Prussian Army also took note, but they took an even more radical step – they adopted steel as a material for their guns. Today, of course, it would take a good deal of mental effort to think of using anything else, but in the 1850s steel was a

The largest calibre British gun ever built, Mallet's 36-inch mortar was built up from rings of wrought iron held together by longitudinal bars. Intended for the Siege of Sebastopol, it was too late for the Crimean War and has stood in Woolwich Arsenal ever since.

An early Portuguese breech-loader with an early design of spring recoil buffer.

material regarded with a good deal of misgiving by engineers. The problem was a technical one – how to produce a block of steel of the requisite size for a gun barrel so that it was sound and of high quality throughout. Numerous attempts had been made, but inevitably, it seemed, the gun would burst during trials due to some undetectable flaw in the steel. Under similar circumstances a wrought-iron gun would split, giving warning of its impending dissolution, but steel guns always disintegrated violently, without warning, and for this reason they were distrusted.

Krupp, the up-and-coming steelmaster of the Ruhr, was determined to produce steel guns, and he performed countless experiments in casting steel into blocks of hitherto unheard-of size, boring them for gun tubes or cutting them up to check for flaws. Eventually he was confident that his gun-casting technique could produce reliable weapons, and after numerous trials between 1844 and 1855 an experimental committee of the Prussian Army proposed some service designs. In 1856 Krupp produced a steel 9-cm calibre field gun to one of these designs, and this was considered so satisfactory that from then on steel was the standard Prussian gun material.

Krupp's other innovation was his breech-loading system. This used a wedge or block of steel sliding transversely in a slot cut in the rear end of the gun tube. Movement was imparted to the block by a slow screw mechanism, and the sealing of the breech end against the escape of gas was done by soft metal 'Broadwell Rings' let into the face of the breech-block where they mated with the face of the breech itself. His system of rifling he changed from time to time, as better ideas suggested themselves; he first used deep grooves with studded projectiles, similar to de Beaulieu's system; then came a system with many shallow grooves and a ribbed lead jacket on the shell, and then the same form of rifling with an expanding soft metal baseplate to the shell.

The question which might be uppermost in the layman's mind at the moment might well be that of why all this insistence on rifled guns? The answer is that there were a number of advantages which accrued from rifling: the principal one was that it allowed the weight of the projectile to be increased. A smoothbore gun used a ball, since this was the only shape which would be stable in flight; an elongated cylinder would have tumbled end-over-end and would have been both highly inaccurate and short-ranging. For any given calibre of gun there was but one sphere which would fit inside, and thus the size of projectile was automatically fixed – the reason why smooth-bore guns were always called 'something-pounder', since if you knew the weight of ball you immediately knew the calibre. But rifling the gun caused the projectile to spin and endowed it with gyroscopic stability; hence it became possible to use an elongated projectile of greater weight and capacity, giving better effects at the target. As a by-product accuracy was improved, and due to the

(*Above and below*) A rifled muzzle-loading 9-inch gun on a high angle mounting. Used for coast defence, it was intended to fire a 360-lb piercing shell so as to drop steeply on to the decks of warships. (*Royal Artillery Institution*)

rifling system used five grooves, the shell being spun by an expanding brass ring at its base.

In fact it fell to the lot of the Americans to be the first nation to go to war with rifled ordnance when the Civil War broke out in 1861. The Union forces had Parrott 12-pounders and the Confederates acquired a variety of ordnance from sympathizers in Europe, including Lancaster oval-bore guns, Whitworth's twisted hexagonal bore and a few Armstrong guns. However, rifled weapons were still comparatively rare, and the majority of the ordnance used on both sides remained smoothbore throughout the war. Prominent among these were the guns cast according to the principle devised by Captain T J Rodman, in which the core of the mould was water-cooled. This ensured that the inner surface of the barrel cooled rapidly and solidified while the remaining metal, slower to cool, placed the inner section under compression due to contraction during the cooling period. This placed a compressive stress on the bore surface and added to the gun's resistance to the explosion of the powder charge, and these Rodman guns were probably the high-water mark of cast smoothbore ordnance. Numbers of 15-inch Rodman guns remained in use as coast-defence guns until superseded by more modern armament in the late 1890s.

In Europe the Austro-Prussian 'Seven Weeks War' of 1866 saw the first clash of armies using significant numbers of rifled weapons, but the Prussians, for all that they won the war, found their artillery sadly lacking. Had the war depended on artillery fire the result would have been a resounding Austrian victory and the subsequent history of Europe might have been totally different, and no sooner was the war over than the Prussians began a ruthless overhaul of their artillery arm. The principal defect was that over the years the artillery had become estranged from the rest of the Army, and tended to act independently without much regard to what anyone else on the battlefield was doing. Too high a proportion of guns were labelled 'reserve', with the intention of holding them back to be thrown in at a decisive point, but the magic of the word 'reserve' caused them to be retained against emergency only and never committed to battle. There had also been a considerable resistance to change among many of the more senior officers; one was so opposed to the idea of rifled ordnance that on his deathbed he gave orders that the salute over his grave was to be fired with smoothbores! The efficiency of rifled guns had been demonstrated in the brief Prussian-Danish War of 1864, where rifled siege guns soon forced the fortress of Düppel, but in spite of this the field artillery held on to their smoothbores in large numbers on account, it was claimed, of their superiority when firing case shot.

As a result, when the Seven Weeks' War was joined, these defects became apparent. The gunners were poorly trained; most of the guns were outranged by their Austrian counterparts;

better sealing of the propellant gas behind the projectile the range was also improved.

In America, too, experiment was afoot. Gun-making had begun there at the time of the War of Independence, and by 1775 both bronze and iron guns were being cast at Philadelphia. France's example of reorganization had been followed in the early 1800s, and in the middle 1850s numerous designs of rifled ordnance were produced and tested. As a result some 300 3-inch guns were produced; their construction was unusual, the barrel being fabricated of layers of wrought iron rolled on a mandrel and welded by heating and hammering into a homogeneous mass. One of the most important names of the period was that of Robert P Parrott who devised a quick and effective system of construction, using a cast-iron barrel reinforced by a wrought-iron breech coil, which gave guns of his design a distinctive outline. His

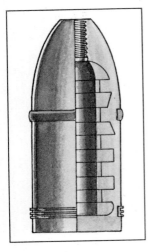

(*Above*) A Russian ring shell, Model 1844. The shell body was built up from rings of iron surrounded by a cast-iron body. Filled with gunpowder, when exploded the shell was shattered, but due to the ring structure fragmentation was under some degree of control.

(*Right*) A section through an early ironclad ship illustrating the system of attaching armour in front of a wood backing.

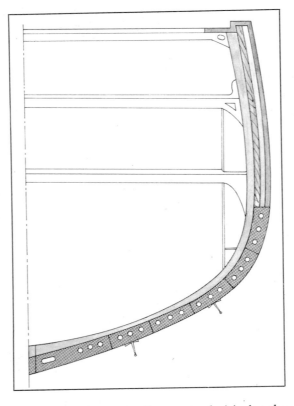

the reserve was never thrown in decisively; the independence of the artillery caused the infantry generals to relegate them to the rear of the column in order to avoid arguments; the guns when brought into action were too widely dispersed to be able to bring concentrated fire to bear on critical targets; and since the loss of the guns was held to be the crowning disgrace, officers were loath to bring their guns forward and put them at risk in the places where they would have done most good. In the majority of cases the artillery was so ineffective that the Austrian guns had perforce to be dealt with by infantry assaults, and when the war was over the infantry were not slow to accuse the artillery of having left them with the hard work—and to a great extent they were right.

Whatever the Prussian artillery may or may not have had, they certainly had an intense pride, and the accusations which flew were sufficient to set in motion a complete overhaul of the arm. The moving spirit was General von Hindersen, an artillery fanatic whose motto was: 'The day contains 24 hours for duty.' He was well aware of many of the defects and had previously demanded some reforms, but these had been denied the necessary funds. He therefore gathered his officers together and they, at their own expense, founded a School of Gunnery, the Government being called upon for nothing more than the cost of the ammunition expended. The foundation of this school was too late to have any effect on the war of 1866, but it was to pay dividends in later years. Now, with the lessons of the war to guide them, the smoothbores were completely replaced with rifled guns, the tactics were radically over-

hauled and a new spirit of enterprise spread throughout the corps.

As a result the Franco-Prussian War of 1870 saw the German artillery pushing forward in advance of their infantry, massing to secure decisive results, and preparing the way for the foot soldiers with accurate and devastating fire. The odium of losing guns had been replaced by the greater odium of failing to do one's duty. 'If the artillery wants to save its guns, it must kill the enemy . . . If it cannot thus save its guns, it saves at least its honour,' said one artillery commander. Guns were never withdrawn from action, even if their ammunition supply ran dry; indeed, in one such case the battery commander made his men climb on to the guns with their carbines and sing 'Die Wacht am Rhein' rather than hide.

After the successful conclusion of the war of 1870, Prinz Kraft zu Hohenlohe-Ingelfingen, who had commanded artillery in the wars of 1864, 1866 and 1870, wrote a series of 'Military Letters' or essays on the tactics and lessons learned during these three wars. They were translated into several languages and they became required reading for every artillery officer in any army with pretensions to efficiency. Prinz Kraft discussed and analysed the organization, equipment, handling, tactics and fire control of artillery in great detail, and it is not too much to say that almost all field artillery tactics from then on have owed something to some observation of Kraft's. He ended his letters with a summary of principles, and the first of these is worthy of repetition:

Of the requirements of a good artillery, the first and principal one may be expressed in three words:
(1) First, it must *hit*; second, *hit*; and third, *hit* . . .
(2) It must be in the right place at the right time . . .
If it fulfils these two requirements it will be able to do anything required in action.

Nobody, before or since, has ever summed up the function and purpose of artillery more succinctly.

The Prussians were now well content with their artillery; understandably the French were less so, and they in their turn began to renew their equipment and rethink their tactics. Their eventual decisions were largely conditioned by their chastening experiences in 1870, but unfortunately when the time came to put them into effect, it was no longer 1870. The British Army, on the other hand, was still searching for the perfect system of equipment before attempting to develop a perfect system of tactics. Armstrong's rifled breech-loaders had eventually been tested in battle, in China and New Zealand by the Army and in Japanese waters by the Navy, and the reports were somewhat conflicting. On the one hand all the users agreed that the range, accuracy and target

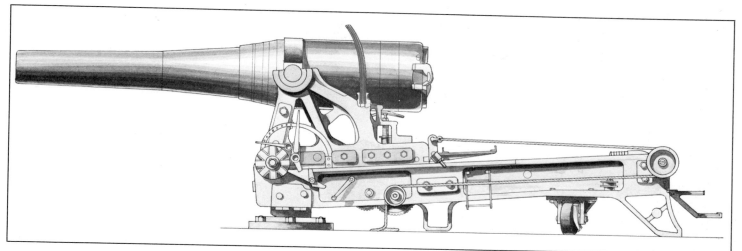

A French 24-cm coast defence gun, Model 1876. Recoil was controlled—to some degree—by compressor plates hung between the slide sides.

Another early recoil control system is seen on this Portuguese coast gun. There is provision in the carriage for a second cylinder, but one appears to have been deemed sufficient.

effect were all that could be desired, but on the other hand it was generally admitted that the breech-closing arrangements were prone to get out of order unless fussed over by a skilled artisan. Any faiiure in maintenance or fault in operation could result in a serious accident; vent pieces cracked and even occasionally blew out; the gas sealing, an arrangement of copper rings in the face of the vent piece, tended to leak, leading to erosion, which led in turn to greater leakage—it was the peterara problem all over again.

In 1863 Joseph Whitworth had perfected his system of rifling in which the bore of the gun was a twisted hexagon, and he requested a trial. The opportunity was taken of performing a comparative trial between the Armstrong RBL Guns, the Whitworth Rifled Muzzle Loader (RML) and a new RML system developed by Armstrong in which the bore was rifled with three deep grooves and the projectile provided with metal studs—more or less the same as that of de Beaulieu and Krupp, though using a lesser number of grooves.

The subsequent trial was most comprehensive—the cost was £35,000, an immense sum to spend on a trial in those days—and in addition the Committee appointed to run the trial sought the

The patent drawing of Captain W Palliser, showing his system of lining old smoothbore guns with a rifled sleeve.

opinions of scores of experts both in the engineering profession and in the artillery. Eventually, on 3 August 1865, the Committee reported overwhelmingly in favour of the Armstrong RML system, since 'the many-grooved system of rifling, with its lead-coated projectile and complicated breech loading arrangement . . . is far inferior for the general purpose of war to both the muzzle loading systems, and has the additional disadvantage of being more expensive both in original cost and in the cost of ammunition . . . Muzzle loading guns can be loaded and worked with perfect ease and abundant rapidity.'

While the Committee had been principally considering field artillery, another factor had appeared on the scene, one which was to swing the scales decisively to the side of the rifled muzzle-loader. The French had begun building ironclad warships, and in doing so they had set off an armament race of incredible proportions. When *La Gloire* was laid down in 1858, she rendered the world's wooden walls obsolete by her very fact of floating. Although the British Admiralty had condemned iron as a constructional material as early as 1840, they were forced to change their tune, and their first ironclad, HMS *Warrior*, was laid down in 1859 in response to the French move. While iron was possessed of constructional advantages, the principal reason for adoption was, of course, its power of resistance to gunfire, and before a suitable design could be drawn up it became necessary to make some trials to find out just how hard the stuff was and how much of it would be needed to keep out hostile projectiles. One of the first discoveries was that iron by itself was of little worth; it had to be supported, or 'backed', by some resilient material which would absorb the shock of the blows. Eventually the British designers settled for 4·5 inches of wrought iron, backed with 18 inches of teakwood and a thin inner iron skin. This, the constructors claimed, was proof against any known weapon at ranges of 400 yards or more. The teak supported the iron against the shock of the projectile's arrival, and, should the shot pierce the plate,

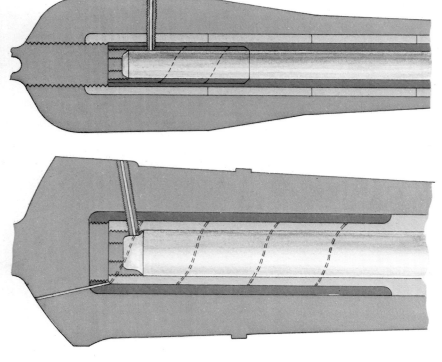

An American 13-inch seacoast mortar of the Civil War period. It introduced the principle of deck attack and was the forerunner of much heavier weapons.

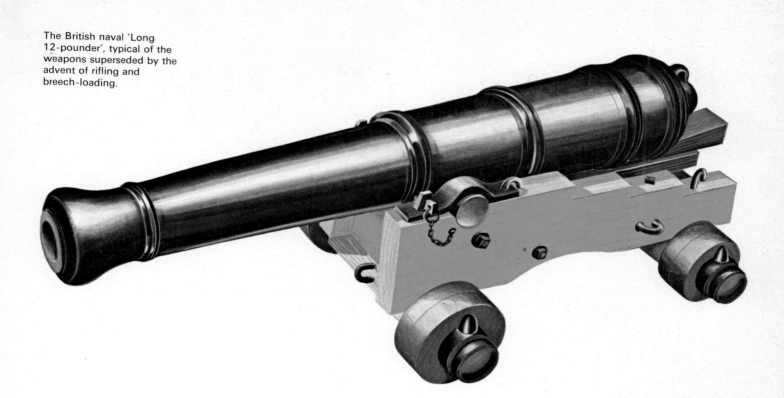

The British naval 'Long 12-pounder', typical of the weapons superseded by the advent of rifling and breech-loading.

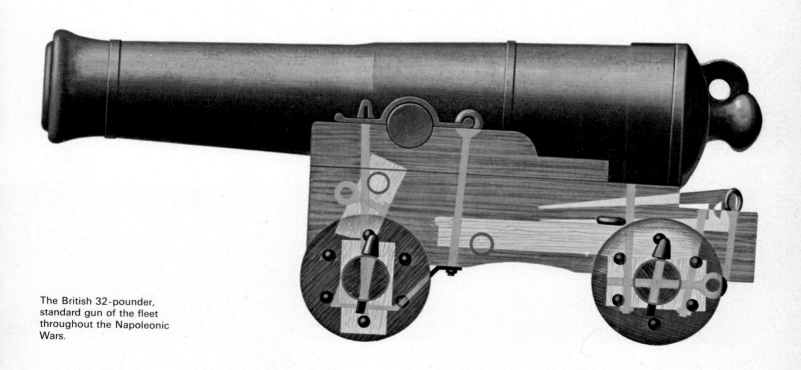

The British 32-pounder, standard gun of the fleet throughout the Napoleonic Wars.

British 7-inch Armstrong
rifled breech-loading gun
on naval slide mounting.
The breech was opened by
revolving the large screw
and lifting the 'vent piece'
out; failure to tighten the
screw after loading could
lead to the vent piece being
blown violently out when
the gun fired.

A boat carriage gun of the American Civil War. For service afloat the gun (*top right*) was stripped from its wheeled carriage and mounted in the bows, while the field carriage was carried in the stern. The vessel was also iron sheathed at the water line in order to afford some measure of protection.

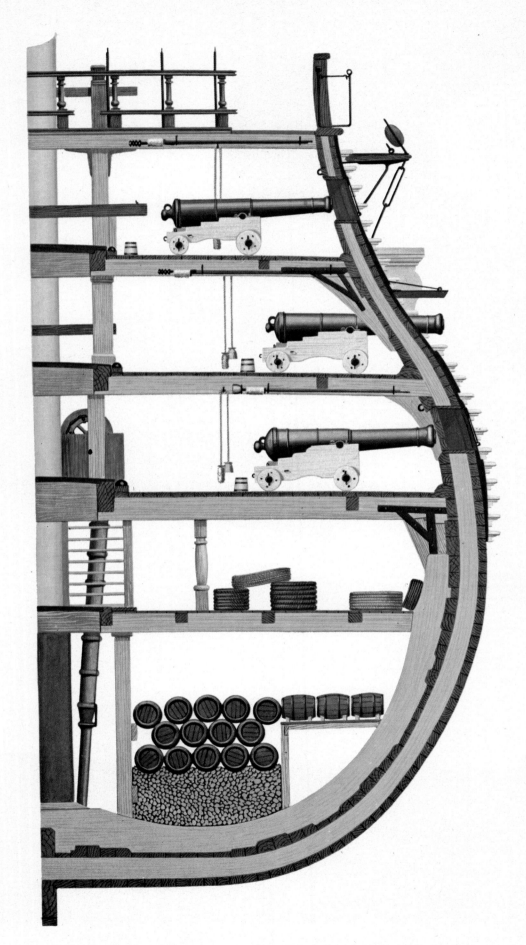

A section through HMS
Victory, showing the
arrangement of gun decks.

(*Top*) An early breech-loader, which exhibits hydraulic recoil buffers alongside a ring cradle in which the gun recoils.

(*Above*) A 9-pounder with the muzzle pierced according to theories advanced by de Beaulieu, an early forerunner of the muzzle brake. (*Royal Artillery Institution*)

also the best method of defeating it; it was a rare example of one body being given the jobs of gamekeeper and poacher at the same time, but in fact they laboured very willingly in both roles and made some fundamental discoveries in both fields of research. One of the first things to become apparent was that the Armstrong gun's breech-closing system was too delicate to withstand the enormous propelling charges which were going to be demanded in order to fling projectiles at sufficient velocity to pierce plate. This conclusion, together with the report of the Armstrong and Whitworth Gun Committee, sealed the fate of the rifled breech-loader in British service, and work began on developing suitable muzzle-loaders for both ships and shore defences.

The launching of *La Gloire* also sparked off an orgy of fortress-building without historical parallel. The British Government, spurred by public concern over the thought of an armoured French fleet just across the Channel, plus the sabre-rattling of Napoleon III, had formed a Royal Commission in 1859 to examine the nation's coast defences. Their report recommended the building of extensive defensive works to the tune of over £11 million, and other nations rapidly followed their example, resulting in the enormous and incredibly expensive fortresses of Heligoland, Kronstadt, Portsmouth, Cork, La Spezia, Antwerp and many others. These works demanded armament on a generous scale (though in the majority of cases they never got it, due to the rising complexity and therefore cost of ordnance as the works progressed) and this again added impetus to the work of the designers.

Between 1862 and 1864 the British Army tried out numerous experimental RML guns, using a variety of rifling systems, as a result of which the 'Woolwich' system, a modification of de Beaulieu's, was adopted. But one advantage of the Armstrong gun, its built-up construction, was retained; the day of the one-piece cast gun was over.

Governments are, of course, always anxious to discover methods of saving money, and the military are always prime targets for the cheese-parers. In the 1860s the search was for a method of using up some of the hundreds of smoothbore guns which were rapidly becoming obsolete in the face of the advance of rifled guns. France, Germany, Austria, the United States, all experimented, and the general solution was to take the old guns and cut rifling grooves in the bores as they stood; the Americans took this so far as to send out artisans with kits of tools to rifle the guns wherever they happened to be mounted. However, a Captain Palliser of the British 18th Hussars came up with the best idea; he suggested boring out the old guns to remove any fissures or erosion marks, and then inserting a rifled steel liner of the requisite calibre. This liner was a tight fit and when in place was expanded and locked firmly in position by firing a heavy proof charge and shot from the gun. The result was, in fact, stronger than

tended to smother its progress. The inner skin prevented teakwood splinters being driven into the ship to act as anti-personnel missiles.

While the constructors drew up their plans and watched their ships grow, the gunners now had their thoughts on the problem. If the ships of the world were going to be ironclad, then both the naval gunners and the land service gunners defending coastlines and dockyards had to think in terms of a weapon capable of beating the new target. The great race of gun versus armour, which is still in progress, had begun. In 1862 the British Army assembled its inevitable committee, this time a 'Special Committee on Iron' to begin a long series of trials and experiments to determine both the best method of utilizing iron armour and

the original smoothbore had been, and several hundred were built and supplied for naval and coast defence use in both Britain and the USA who also adopted the Palliser system.

Having got the weapons organized, the next problem was to produce a projectile which would defeat armour, and here two distinct schools of thought appeared. The first school, which found many supporters in America, since it was particularly well suited to their large smoothbore guns, was the 'racking' party. They contended that the best method of attack was to hammer the target with heavy blows so as to 'rack' or strain the whole structure so as to bring about the collapse of the armour and reveal the target which the armour had been protecting. On the other side was the 'punching' school, who insisted that the

proper method was to drive the projectile through the plate so that both it and the fragments of armour it tore or pushed out in its passage would cause injury and damage behind the armour. In short, the rackers were out to destroy the armour, while the punchers were out to disable the men and equipment *behind* the armour.

Argument waxed loud and long, but eventually, so far as the British were concerned, it was settled in 1863 by a paper written by a Lieutenant W H Noble, MA, a member of the Ordnance Select Committee, who instanced a trial he had just conducted and which had confounded a number of arguers. A 68-pounder smoothbore and a 7-inch Armstrong gun firing a 200-lb shot had been fired at 4·5-inch plate backed by 18 inches of teak. The 68-pounders had penetrated

A variety of projectiles developed for naval use, their prime intention being to do as much damage as possible to the enemy's masts and rigging. These examples include chain, bar and split shot.

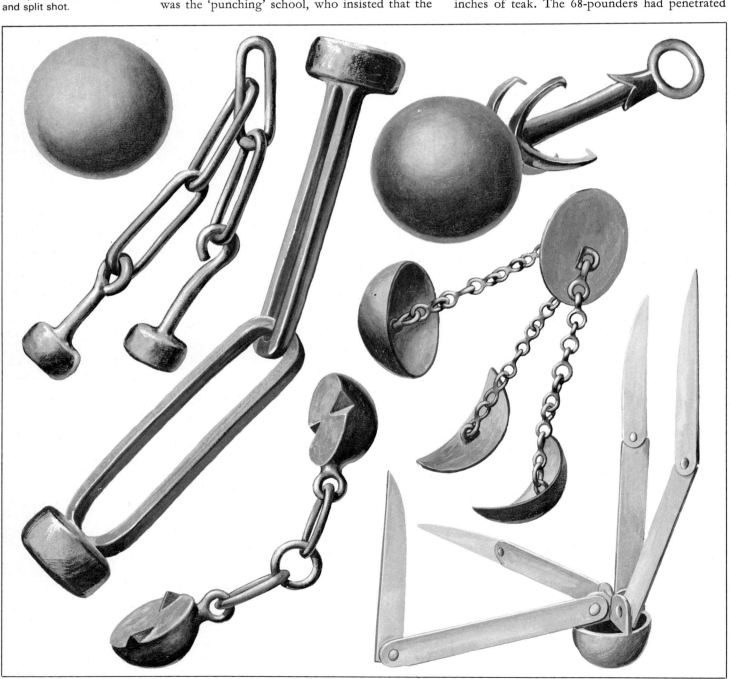

the target, while the 200-pounder had made hardly any impression on it. Noble showed that the answer lay in the relative velocities of the two projectiles: the 68-pounder had been moving at 1,425 feet per second, when it struck, while the 200-pounder was loafing along at a mere 780 feet per second. On the face of it, the 200-pounder, with 156,000 foot-pounds of energy, should have out-performed the 68-pounder with 96,900, but the low velocity of the heavier projectile allowed the plate to deform and resist the blow, whereas the higher velocity of the 68-lb shot tore through the plate before it could begin to absorb the blow. It took a lot more argument and several pages of mathematics to completely convince the sceptics, but the racking school had to give way to the punchers. As Noble said in his paper, 'What is

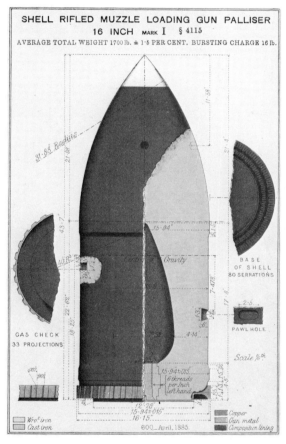

wanted is *velocity*; if you sacrifice it to weight you will only be able to keep knocking at the door without entering.'

One of the believers in penetration was the same Captain Palliser who had devised the system of relining smoothbore guns. While most people, mesmerized by the racking theory, were making tests with flat-headed or round-headed shot, Palliser designed a pointed projectile. In order to ensure its penetrating power, he devised a method of casting the shell nose down in a water-cooled iron mould to form the point, and a normal sand mould for the remainder of the shell body. This rapid chill to the point endowed the cast iron with incredible hardness and the result was a highly

efficient projectile. It was rapidly adopted through-the world as the standard method of penetrating armour, though on the Continent there was still a good deal of experimenting with steel projectiles. In 1879, during the war between Peru and Chile, the Chilean *Almirante Cochrane* fired a 9-inch Palliser shell which pierced the turret of the Peruvian *Huascar*, passing through 5·5 inches of iron plate, 13 inches of teak and a half-inch inner skin to kill most of the turret crew and completely wreck the gun within.

Prior to the 1850s little of importance had occurred in naval ordnance save for the introduction and brief glory of the 'Carronade'. The origin of this celebrated weapon can be traced back to Benjamin Robins, who published a treatise in 1747 arguing for a better distribution of metal in cannon design. He considered that a large-calibre gun was necessary to damage enemy ships, but that the barrel should be made as lightly as possible commensurate with the pressure it had to sustain. While this pamphlet was generally well received, nothing more came of it until in December 1778 the Carron Company, a Scottish iron-founders, built some light guns with which to arm their sailing ships. Once the news got abroad, others began to demand the new weapon, and the Carron Company found themselves in the ordnance business, making their new carronade in quantity.

The new weapon was a short and light gun of large calibre, using a standard cannon shot with a small charge to give a short range. One of the virtues of the carronade was that it was carefully made so that the calibre of the gun and the diameter of the ball were very close—to use the technical expression, the shot had small windage. Obviously a muzzle-loading projectile could not be a tight fit in the gun bore, since air had to escape as it was loaded, but since the invention of the gun, windage had never been very closely considered; it was generally about a quarter of an inch, irrespective of the calibre of the gun. Using a smaller windage resulted in greater efficiency, since less of the propelling gas was blown past the shot, and it also greatly improved the accuracy.

The carronade was rapidly and widely adopted in the Royal Navy since it particularly suited the British style of naval action—to get close to an enemy and let fly with a disabling broadside at short range. It was less quickly adopted by other navies whose doctrine was rather to stand off and use long guns to disable their opponent at longer ranges. This fundamental difference was argued to and fro for many years among sailors; it was left to the American Navy to develop a compromise weapon and finally bring about the downfall of the carronade. In the War of 1812 the American frigates used the 'Columbiad', a weapon halfway between carronade and long gun, firing a heavy shot to a greater range than the carronade. During the war the Americans demonstrated that provided a captain had the skill to keep away from an English ship and fight at a

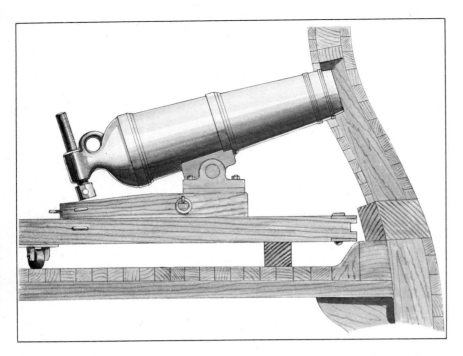

A 32-pounder Carronade on ship mounting, showing the light construction of the class of gun.

had tended to stultify innovation in the Royal Navy. It was slowly becoming apparent that solid shot was no longer the complete master of the ship–no ship was sunk at Trafalgar, for example–and the French began to explore the possibility of using shell-firing guns in order to produce a more lethal and damaging effect. As early as 1798 a series of experiments were carried out at Meudon, firing 24-pounder and 36-pounder shells at ranges of 400 and 600 yards which gave promising results. Finally, in 1822, General Paixhans published *Nouvelle Force Maritime et Artillerie* in which he advocated a system of naval gunnery based on standardization of calibre and the use of shell guns. He admitted that neither idea was new or revolutionary, but he expounded his arguments more eloquently than had been done previously. His proposal was for a fleet of steam vessels armed with shell guns based on the calibre of the existing 36-pounder gun. It took some time for officialdom to accept his ideas *in toto*; the principle of 'unity of calibre' was accepted first, the French Navy adopting a standard calibre based on their existing 30-pounder, constructing a number of models of gun of varying size and weight to suit mounting upon different decks.

Paixhans then designed a shell gun, an 80-pounder of 22-cm calibre, short, undecorated, utilitarian, using a small charge. At a demonstration against a moored frigate the damage wrought by this gun was tremendous; but in the way of all such bodies, a committee appointed to examine Paixhan's propositions took its time over innumerable tests before finally approving the idea in 1837.

Such a move could not be ignored; the British Navy rapidly looked into the matter and began to reduce the number of calibres of ordnance used in ships and then to develop shells for them. Other nations followed suit, and by the time of the

range of his own choosing, he could do pretty much as he pleased. The lessons learned on the Great Lakes, when the Americans severely punished the Royal Navy was reinforced by chance when an American–the *Essex*, armed with carronades–met the British *Phoebe* armed with long guns. The *Phoebe* chose the range and soon disabled the *Essex*, and from then on the carronade's day was over.

The long superiority of British gunnery in naval battles against the French had given the stimulus to some original thinking on their part while it

An American 'Columbiad' on a coast defence mounting. The method of adjusting elevation was unusual, as was the heavy cast-iron central pivot. The carriage wheels were provided with holes to allow the gun and carriage to be levered back into place after recoiling along the slide.

Crimean War the shell gun was an accepted feature in every navy.

Paixhan's other thesis, which he introduced along with his artillery suggestions, was to protect ships from enemy shell by iron plates, and as we have seen this blossomed into the ironclad vessel and set in train any number of consequent actions. But so far as naval vessels went the matter did not simply lead to hanging teak and iron plates all over a ship. One odd feature which stemmed from it was the sudden upsurge of interest in the ram as an instrument of destruction; by providing the ship's prow with a heavy steel ram the whole vessel could be used as a weapon, and this theory held sway for several years, until the increased power of artillery led to the wider separation of fleets. It was helped to survive by the action at Lissa between the Austrian and Italian navies, when the Austrians used rams to good effect, and conditioned a lot of naval thinking thereby.

The advocacy of the ram brought about an increase in the amount of time that a vessel would be head-on to an opponent, and this in turn brought new light to bear on another question. For many years it had been apparent that the broadside action had been losing its prominence, and numerous modifications had been tried in endeavours to increase the arcs of fire from the old broadside mounting of the gun on its truck carriage. Now the ram increased the demand for guns which could fire forward; one solution was to shape the ship's side to allow one or two guns to shoot ahead, but this merely ruined the vessel's lines without adding much to the fighting ability.

Another point raised itself; if the guns were to be protected by armour, then scattering them all over the ship would demand a considerable weight of iron, and thus there was a good case for concentrating the guns into a 'battery' which could be more easily protected with the minimum weight of armour.

All these conflicting arguments led to a number of solutions of varying effectiveness and duration. The central battery was an early contender, an armoured box amidships with ports both broadside and fore and aft. The guns within could be moved from side to side or to command ahead or astern as the situation demanded. What sort of pandemonium must have ensued inside the box during a heated action with guns being trundled back and forth was, fortunately, never put to the test of battle. Various modifications to the central battery to try and overcome this disadvantage were tried, but eventually the weight of opinion was that the guns should be able to develop their greatest ability in the fore-and-aft mode, with the improved manœuvrability of a twin-screw steam vessel giving them the opportunity to do most damage.

The next step was the 'sponson', a semicircular platform projecting from the ship's side and mounting a gun so that it could cover 180 degrees from ahead to astern, a system widely adopted by the French.

Both the central battery and the sponson demanded some more convenient form of gun-mounting than the truck carriage, and the recoil slide slowly replaced it. Here the gun was fitted into what amounted to a truck-less truck carriage; this in turn rode on an inclined platform secured to the ship's side to form its traversing pivot and with its inboard end supported on iron trucks running on iron 'racers' or traversing arcs let into the deck. The gun and carriage recoiled up the slide, controlled to some extent by friction and breeching ropes. It was then loaded and run out again by a combination of tackles and, if the ship was at the right heel, gravity. This was later improved by the addition of the 'compressor', interleaved plates, some attached to the gun carriage and some to the slide, and which could be placed in compression by a screw-jack. On recoil the gun-carriage plates were dragged through the grip of the slide plates; for running out, a quick release opened the 'pack' of plates so that the gun carriage could run free.

(Below) An Elswick Ordnance Company design of 'broadside mounting' for a 6-inch gun. This is the 'Vavasseur' type of mounting in which the gun recoils up an inclined plane and returns by gravity.

(Bottom) A British 11-inch of 25 tons, rifled muzzle-loader, mounted on a slide and carriage of the HMS *Téméraire* type. The side-arms in the foreground include a 'worm' for removing unfired cartridges.

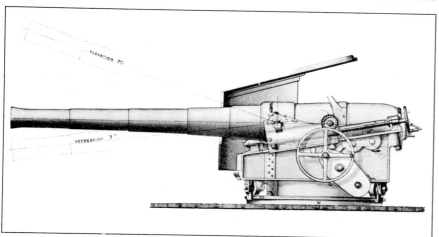

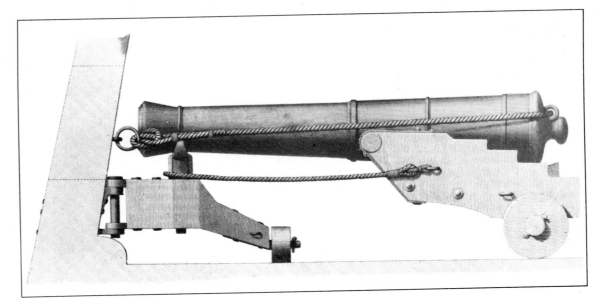

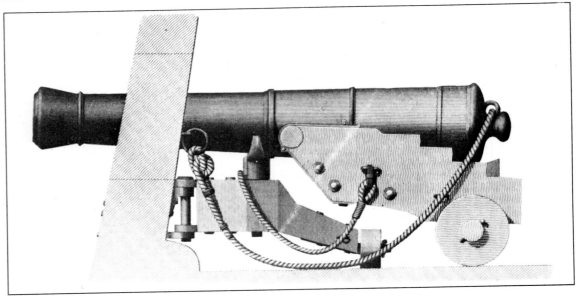

In spite of a variety of central battery designs, none were entirely satisfactory, and to attain the desirable ideal of being able to fire in any direction it became necessary to pile on ordnance, even to the extent of having two-storied batteries. With the rapidly increasing size and weight of guns which were having to be developed to keep pace with armour, the great weight high in the ship became a potential hazard and a better solution had to be found.

It had, in fact, been available for some time, and its supporters had been urging its adoption in place of central batteries for a number of years: the centre-line turret was their article of faith. The invention of the turret has been claimed in various quarters, but it seems that the credit lies with Captain Cowper Coles, RN. In 1855 in the Sea of Azov, he built a raft to carry a heavy gun and with it shelled the Russian stores depot at Taganrog, this being the only way to approach within range due to the shallow waters. This device took the Admiralty's fancy, and Coles was

ordered home to advise on the construction of a somewhat more elegant version. In this he speci-fied the need for armoured protection of the gun, and he designed a shallow-draught vessel equipped with heavy guns mounted behind a fixed shield. Shortly thereafter he conceived the idea of making the gun and its shield revolve in what he termed a 'cupola', and the turret was born. In 1861 the Danish Government enlisted Cole's advice and built the *Rolf Krake*, a turret gunboat mounting four 8-inch guns in two armoured turrets. Their example was rapidly followed by Prussia, Italy, Brazil and Russia and the turret ship was sud-denly popular.

However there were divergencies of opinion as to how the turrets ought to be disposed and what sort of ship ought to be built to carry them; the controversy had a fatal result in one instance. The British Admiralty had a ship built—the *Monarch*—with Cole's turrets: it was a high-freeboard ocean-going ship which performed well. But Coles disagreed with the design, and insisted on having

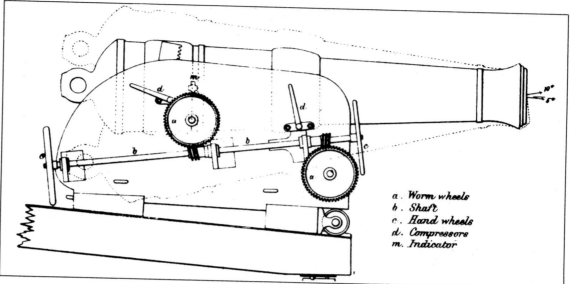

A contemporary drawing showing Shaw's muzzle-pivoting carriage; this elevated the gun about an imaginary pivot at the muzzle, thus allowing the use of the smallest possible port in the ship's side.

a. Worm wheels
b. Shaft
c. Hand wheels
d. Compressors
m. Indicator

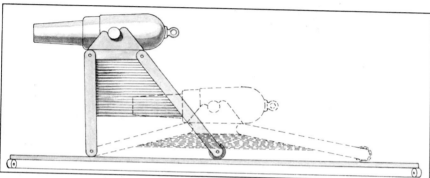

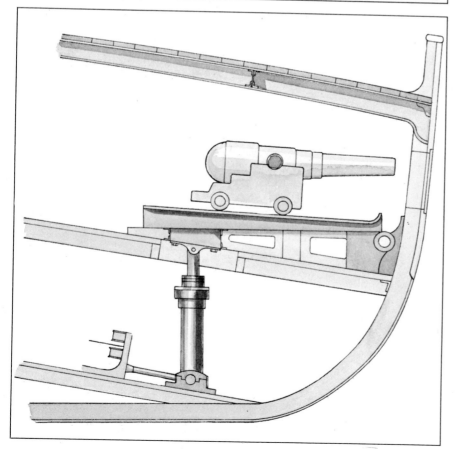

a second ship built, the *Captain* with a low freeboard; by some draughtsman's error the freeboard was even lower than that planned, as a consequence of which it capsized in a storm taking the lives of Captain Coles himself and most of the crew.

The problem at the heart of the matter was reconciling turrets with a full-rigged ship, and the loss of the *Captain* finally showed the dangers inherent in such an arrangement. Even before this unhappy event the Admiralty had begun construction of a mastless turret ship, in which decision the famous American *Monitor* and its offspring had considerable influence. The loss of the *Captain* also led to a good deal of fundamental study of stability, as a result of which design now moved away from the low freeboard, as exemplified by *Monitor* to a higher freeboard giving better sea qualities.

While the turret gun was now accepted, some doubts remained. It seemed ridiculous to require an enormous vessel simply to carry four guns into battle, albeit these guns were by now powerful enough to deal with anything they could hope to meet. The development of light and fast torpedo-boats led to a proliferation of minor armament, from 6-pounders to machine-guns, and eventually the design of warships began to crystallize into the form in which it was to remain: heavy guns in turrets at each end, supported by broadside batteries between them.

All this, of course, led to advances in the methods of mounting the guns. The first turrets were hand operated, by sailors cranking handles, but this was soon abandoned in favour of machinery, first operated by steam and later by hydraulic power. Once power was available the turrets and their guns could be increased in size and weight to such a degree that the *Inflexible* of 1874 had two turrets each weighing 750 tons and mounting two 12·5-inch rifled muzzle-loading guns. The form of naval armament was settled; all that now remained was mechanical improvement.

CHAPTER 4

The Age of Invention

With the improvement in ordnance well under way it became necessary to invent means of measuring performance with greater exactness than had hitherto been possible, or indeed considered necessary. Now, with the problem of armour before them the gun-makers and gunners had to be able to make precise comparisons between different charges, different weights of projectile, different lengths of gun, and determine just how each of these and many other factors were related to the eventual performance of the weapon. It was no longer sufficient to merely reach out as far as possible or penetrate some simple target.

Professor Wheatstone began experiments in determining the velocity of projectiles by electric means as early as 1840, and in subsequent years many other experimenters were drawn to this problem. The first practical apparatus was developed by Major Navez of the Belgian artillery in 1848, though it was not until about 1855 that he perfected it and saw it taken into use by various countries. The British Army obtained one of his devices at about this time and it was installed at Woolwich to determine velocities. Navez's instrument was literally an electric variation of the ballistic pendulum; a pendulum was held at one extremity of its swing by an electromagnet. A wire stretched in front of the gun was cut by the passage of the shot, breaking the flow of current and allowing the pendulum to begin its swing. A second wire, at a measured distance, caused a second electromagnet to arrest the movement of a light index plate carried by the pendulum, so that the passage of the shot was recorded as a portion of the pendulum's arc of movement. This device gave much more accurate results than had previously been possible and, even though it was later described as being 'undoubtably a great improvement . . . for rough practical work . . .' it allowed much basic research into the value of improved types of powder, the effect of various lengths of barrel and twists of rifling, systems of ignition and similar modifications.

Navez's instrument was universally superseded by a much improved device invented by another Belgian, Captain-Commandant P le Boulenge. This also relied on electromagnetism, dropping a bar of metal by means of the first cut wire and incising a mark on it when the shot cut a second wire. The Boulenge Chronograph, improved from time to time, became the standard method throughout the world of measuring velocity and has remained in use to this day, though it is now being abandoned in favour of radar techniques.

But both these instruments were capable of measuring only one interval of time, and while they provided a mean of the velocity across a measured distance, they were of little assistance in determining the performance of the projectile throughout its flight. One of the greatest problems occupying the minds of experimenters and ballisticians at that time was the calculation of a formula which would account for the retardation of the shot by the air through which it was passing, and for this to be determined it was necessary to have an instrument which could measure a number of time intervals in succession, so that the shot's flight could be split up into stages and examined. The necessary machine was finally developed by Frances Bashforth, a Bachelor of Divinity who had become the Professor of Applied Mathematics to the Ordnance College at Woolwich. His machine used a number of discs on a common shaft, revolving at a regulated speed; each disc was covered with lamp-black and provided with a scriber operated by the usual electromagnet connected to a thin screen of wire. These screens were placed at measured distances in front of the gun, and the passage of the shot, breaking the screens, marked each disc in turn, thus giving an accurate measurement of the time elapsed between each screen. From this, Bashforth was able to produce the first accurate figures relating to the performance of various types of projectile during their flight, illustrating the fall in velocity and allowing a formula to be calculated so that ballisticians could now predict the expected performance of a design much more closely than had previously been possible.

Now that the flight of the projectile from the gun's muzzle had been determined, the only sphere remaining was what actually happened inside the gun. Rodman, the American designer, had developed a form of pressure gauge which could be screwed into specially prepared holes in an experimental barrel and which allowed some determination of the pressures at various points in the bore, and this was later improved in Britain into the 'crusher gauge'. This was a cylinder of steel screwed into the gun and containing a steel plunger; the plunger bore against a carefully manufactured and measured slug of copper of known malleability. The rush of gas within the gun would drive the plunger outward and compress the copper, and from measurement of the copper before and after crushing the pressure in the gun could be deduced. Finally, Andrew Noble, having left the Army and taken up employment with a gunmaker, developed his 'chronoscope'; this was a similar device to Bashforth's chronograph but actuated by switches screwed into the gun bore in a similar manner to the crusher gauge. As the shell passed up the bore it tripped the switches in turn, allowing the passages to be divided into small sections and the actual velocity of the projectile deduced. The combination of knowledge of bore pressure and projectile movement allowed the experimenters to make new discoveries into the manner of burning of the propelling charge which led to much improvement in gunpowder.

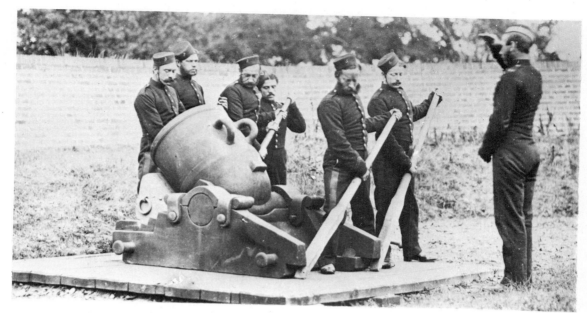

Drill evolutions with a 13-inch mortar. (*Royal Artillery Institution*)

A 10-inch howitzer being loaded. (*Royal Artillery Institution*)

Running up, the operation of bringing a muzzle-loading gun into position. (*Royal Artillery Institution*)

'Prepare for Action' is ordered, and the detachment ensure that all is in working order before commencing fire. (*Royal Artillery Institution*)

Loading with grape-shot. (*Royal Artillery Institution*)

Until the arrival of the Armstrong guns the powder had changed but little since the introduction of corned powder. The actual percentages of the various components had altered over the years, reflecting both the lowered price of saltpetre and sulphur and the increase in the strength of guns, but it was still a fine-grained and fast-burning explosive. With the Armstrong guns it was desirable to provide a powder which burned somewhat more slowly in order to develop its power more gradually, in view of the initial check to movement due to the lead-coated shell having to cut into the rifling, and a form of powder known as RLG (for Rifle Large Grain) was introduced. While essentially the same gunpowder as before it was formed into, as the name suggests, large grains, which had the effect of inhibiting the rate of burning at first and thus delivering power in a more gradual fashion. With the introduction of longer and heavier RML guns even RLG powder was too

violent and numerous experimental compositions were tried out before a suitable propellant was found. Rodman, that indefatigable experimenter, argued that if the charcoal were less purified this might slow down the rate of burning, and he developed 'Cocoa Powder', so-called from its colour; it was made from the normal ingredients, but the charcoal was only partially charred, giving the brown colour. A similar composition–'Brown Powder'–was also developed in Germany. The British developed 'P Powder', pressing standard gunpowder into large grains with higher density and thus obtaining the required rate of burning, and followed this was 'P2', even larger and more dense for the heaviest guns. But for all their success it was apparent that simply making bigger lumps was not the whole answer, and after more trials the technique of moulding fine powder into special shapes was perfected. These were generally called 'Prism' powders, from the shape of the

Loading a coast gun on its traversing platform. (*Royal Artillery Institution*)

Loading with shot. (*Royal Artillery Institution*)

Elevating the gun. (*Royal Artillery Institution*)

Manhandling the trail in order to point the gun. (*Royal Artillery Institution*)

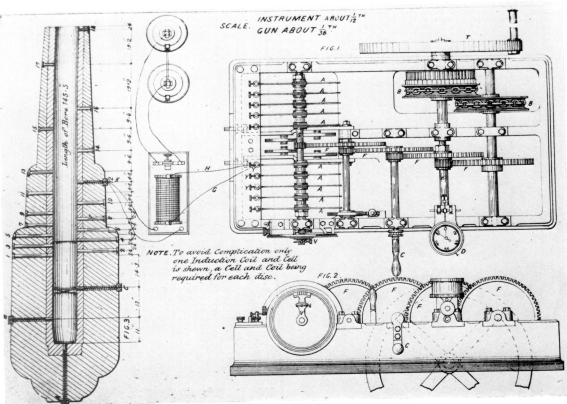

An early ballistic instrument; Lieutenant Noble's Chronoscope, by means of which the passage of the projectile up a gun barrel could be accurately measured for the first time.

moulded grain, and with either solid or perforated grains a great deal of control over the burning rate was possible. By the middle 1880s the British artillery was equipped with a comprehensive range of gunpowders to suit every possible type of weapon.

SBC (Slow Burning Cocoa) for 16·25-in and 13·5-in BL guns

EXE (Extra Experimental) for 6-in BL and 12·5-in RML

Prism 1 Black for 17·72-in, 12·5-in and 10·4-in RML

Prism 2 Black also for 17·72-in and 12·5-in RML

Prism 1 Brown for 16-in RML and a variety of BL guns

P (Pebble) for reduced charges in RML guns

P2 (Pebble 2) for medium calibre RML and BL guns

SP (Selected Pebble) for some of the early BL guns

RLG 2 (Rifle Large Grain 2) for small-calibre RML

RLG 3 for use in India, its formulation being better suited to use in hot climates

RLG 4 for use in small RML and BL guns

RFG (Rifle Fine Grain) for use in short-barrelled howitzers.

However, whatever shape, colour or form gunpowder came in, it was still gunpowder, and it was still prone to the defects which had dogged it since its first invention. It still gave forth clouds of white smoke which revealed the gun's position as soon as the first shot was fired – and went a long way to providing the 'fog of war' which soon enveloped the battlefield – it left quantities of solid fouling in the bore which, unless regularly removed, could soon build up to a point where insertion of a shell became impossible; it was highly susceptible to damp; and of course, it was highly dangerous, responding instantly and fatally to any friction, spark or flame. With the advance of chemistry in

An ammunition carrier belonging to the Austro-Hungarian mountain artillery, carrying eight shrapnel shells and their cartridge for the 90-mm gun, Model 1875.

the nineteenth century it was to be expected that sooner or later somebody would look more closely at the problem of pushing projectiles out of guns and produce some better substance for doing it.

The first modern explosive was produced by Schönbein, a Professor of Chemistry at the University of Basle, in 1846. This was later to be known as guncotton, produced by the action of nitric and sulphuric acids on cotton. Schönbein kept his process a secret and attempted to interest a number of governments, but in the same year another academic, Professor Bottger of Frankfurt-am-Main also hit upon the same substance, and

the two professors joined forces on a profit-sharing basis. Manufacture began in France and Britain in 1847, but after a series of disastrous explosions of unexplained cause, the manufacture of guncotton was prohibited and remained so for another 16 years. It is known now that the trouble was due to insufficient purification of the guncotton, leading to spontaneous combustion, but at the time the whole thing was a mystery and most people were content to let it stay one. In Austria Baron von Lenk spent many years studying guncotton and eventually solved the problem, but even so it was still somewhat unsafe and he was unable to persuade the Austrian Government to accept it for service use. In spite of many trials it was never successful as a propellant, largely due to the fact that it was porous, and under the high pressure developed in the gun chamber the ignition, instead of progressing regularly on the surface as hoped, penetrated the substance and led to violent detonation instead of a controlled explosion.

It was the Prussian artillery who first received the benefit of a modern smokeless propellant with the introduction of Major Schultz's powder in 1865. Schultz Powder was a nitro-lignose impregnated with saltpetre, and it promised to counter every drawback of gunpowder: it was, to a degree, smokeless; it was much less susceptible to damp; it left practically no fouling; and it was relatively insensitive to shock or friction. However, though it was successful in smoothbore ordnance, it was still too rapid and violent for rifled guns. The first smokeless powder suitable for rifled weapons came from France, with Vielle's 'Poudre B' (B for General Boulanger). This solved the problem of rapidity of burning by gelatinizing the powder into a hard impervious substance which could only burn on its surface, so that control of burning could easily be achieved by control of the dimensions of the powder grains.

In Britain the Government Chemist, Sir Frederick Abel, assisted by Sir James Dewar and Dr Kellner, developed a similar substance using nitro-glycerine and nitro-cellulose, to produce 'Cordite', so named from its physical form of long cords. In addition to its virtues of smokelessness and controllability, this, like other modern powders, was far more powerful than gunpowder because of its greater heat content. On the other hand this greater heat led to the more rapid erosion of gun barrels, since the temperature of the explosion was higher than the melting-point of steel. In the United States 'Pyro Powder' a variety of nitro-lignose was developed, followed by a nitro-cellulose gelatinized powder, and by the end of the nineteenth century every nation had some form of smokeless powder in use.

The next field for development lay in the projectiles themselves. Until the arrival of rifled ordnance there were basically three types of projectile – shot, shell and shrapnel. The shot was a plain iron ball; the shell was a hollow iron ball with a filling of gunpowder; and the shrapnel we have

The British 4-inch jointed mountain howitzer; in order to reduce the load on the mules, the piece comes apart and is secured for firing by a connection forming part of the trunnion section.

already discussed. To go with his rifled gun Armstrong produced the 'Segment Shell', a hollow shell built up from segments of cast iron with a cast-iron jacket, the whole coated with lead which both held it together and bit into the rifling to provide the spin. Inside went a cloth bag filled with gunpowder, and into the nose went a time fuze, another Armstrong invention. This used a ring of gunpowder as a timing device; the ring was positioned according to a scale of ranges engraved on the fuze body, and this positioning controlled the distance between the beginning of the train of powder and a vent leading to the fuze's 'magazine', a small compartment holding gunpowder. On firing, a sensitive detonator in the fuze was jerked down by the shock of discharge so that it struck on a fixed needle, and the resulting flash ignited the powder train to burn until it reached the entrance to the magazine, whereupon the magazine was ignited and the subsequent small explosion ignited the contents of the shell. While these fuzes were moderately successful, the detonator was a source of constant trouble, either failing to fire at all or firing during normal handling due to the use of extremely sensitive compositions. The segment shell itself was most effective, though it never lived up to Armstrong's avowed hope of replacing the shrapnel shell. Indeed, it was a notable 'first' in the ammunition field, pre-dating the current fashion for controlled fragmentation by some 100 years.

The shrapnel shell, of course, had to be redesigned with the arrival of rifled guns. Colonel Boxer, Superintendent of the Royal Laboratory, the oddly named ammunition-manufacturing section of Woolwich Arsenal, had already redesigned Shrapnel's original spherical shell twice in order

to make it more safe and reliable and had, in addition, designed an excellent powder-burning time fuze to go with it. Now he set to work for a third attempt, and produced a projectile which while retaining Shrapnel's principle, was totally new from stem to stern and which remained the model for shrapnel of every nation for the rest of shrapnel's days. A small gunpowder charge at the base of the shell was surmounted by a steel plate and a charging of musket balls. A central tube communicated between the nose of the shell and this powder chamber. A time fuze in the nose completed the assembly. When the fuze operated, its flash passed down the central tube to explode the gunpowder. This blew the musket balls forward, forcing off the lightly pinned head section of the shell and allowing the musket balls to disperse. They were further assisted in this dispersion by the spinning of the projectile which now distributed the balls in a predictable and highly lethal cone.

One of the drawbacks of the rifled gun was that there was no longer any windage – the excess of bore diameter over shell diameter which was, of course, necessary to allow the spherical shell to be loaded. Hence no longer did the flash of the exploding cartridge 'wash over' the shell to ignite the fuze, and it was now necessary for Boxer to redesign his fuze as well, using a detonator and needle in much the same fashion as Armstrong. But this was not much more successful than Armstrong's, and with the arrival of the rifled muzzle-loader, and the concomitant reappearance of windage, the older patterns of cartridge-ignited fuze were reverted to with some sighs of relief, until further basic research had been done into the vagaries of detonators.

Other designs of shell also made their appearance, for shrapnel and segment were purely anti-personnel devices. For demolishing defences in the field, 'Common' shell were used, a simple hollow projectile containing gunpowder. These were originally made without fuzes, since the shock of arrival of the shell generated enough friction within to cause ignition of the powder; but the shock of firing from the new and powerful guns was more likely to upset the contents at the wrong end of the trajectory, so that the powder

(*Above*) The coast defence battery at Castle Cornet, Jersey, in the 1860s. (*Royal Artillery Institution*)

(*Right*) For short-range work, case shot was preferred to shrapnel. A 3½-pound chilled iron shot was a formidable missile by itself, let alone in company with forty-odd others.

was enclosed in a cloth bag to insulate it from the rough interior of the shell and a fuze fitted to attend to ignition. Since the shell was to function on impact with the target, the fuze was designed to operate on striking, but, strange as it may seem, this was harder than time functioning to achieve reliably and it took several attempts before a safe and reliable impact fuze was produced.

For piercing armour a plain pointed shot was the accepted form, but it was not long before the suggestion arose of placing a small explosive charge in a cavity in the shot so that after penetrating the armour the shell—as it now became—would detonate and thus do even more damage. This was easier said than done, since putting a cavity in the shell reduced the mass and therefore the striking energy, and also the violent impact against the plate generally set off the shell's contents before the required time. This not only negated the intention but, of course, ruined the shell's penetrative ability. Gradually, by enclosing the charge in a bag and developing a suitable fuze to screw into the bottom of the shell, the problem was solved, but it remained a constant irritant throughout the life of armour-piercing shells; there was no cut-and-dried formula, the whole design had to be sweated through every time a new shell was developed.

As with propelling charges, so with shell fillings; the advent of new explosives led to a variety of experimental shells filled with everything from blasting gelatine to guncotton, by way of some weird and inherently dangerous variations. The selection of a high explosive to go into a shell is fraught with conflicting demands. It had to be violent enough to shatter the shell into lethal fragments, but inert enough to suffer being fired

from a gun. It had, at the same time, to be sensitive enough to be detonated easily by the small impulse given by the operation of the fuze. It had to be cheap and simple to put into the shell; it must not set up chemical reactions with the metal of the shell body; it must remain stable so as to stand storage for long periods of time without deteriorating. One after another the high explosives were tested and found wanting in one or more of these desiderata; probably the only reasonable solution was guncotton, though the problem of detonating

SHOT R.M.L. CASE SPECIAL 9-INCH MARK VI | L

WITH 3LB 9½OZS CHILLED IRON SHOT

SCALE ⅕   § 8425

AVERAGE TOTAL WEIGHT 256 lb.

it efficiently was far from simple. In most countries the experimenters gave up trying what was available and took to searching for some explosive which was more amenable to shell filling. In others the whole problem was turned upside-down and the experimenters took to developing weapons which could fire shells filled with the most violent explosives.

In this latter category Mr Mefford of Chicago deserves a mention for developing a giant air-gun; using compressed air as a launching medium meant that the initial movement of the shell would be relatively gentle, building up, in a sufficiently long gun, to a worthwhile velocity. Such a shell could be filled with the most powerful explosives without having to worry unduly about the sensitivity aspect. Mefford built his first gun in 1883 and it was set up and fired at Fort Hamilton New York in January 1884. One of the observers was a Lieutenant Edward Louis Zalinski of the US Artillery. Mefford, after the trial, decided to make

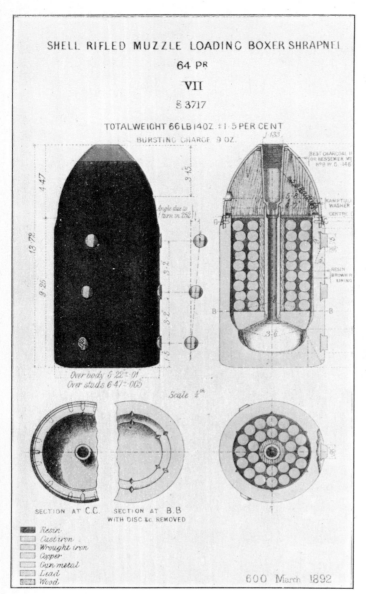

SHELL RIFLED MUZZLE LOADING BOXER SHRAPNEL

64 PR

VII

§ 3717

TOTAL WEIGHT 66 LB 14 OZ. ± 1·5 PER CENT.
BURSTING CHARGE 9 OZ.

SECTION AT C.C.   SECTION AT B.B
WITH DISC &c. REMOVED

Resin
Cast iron
Wrought iron
Copper
Gun metal
Lead
Wood

600 March 1892

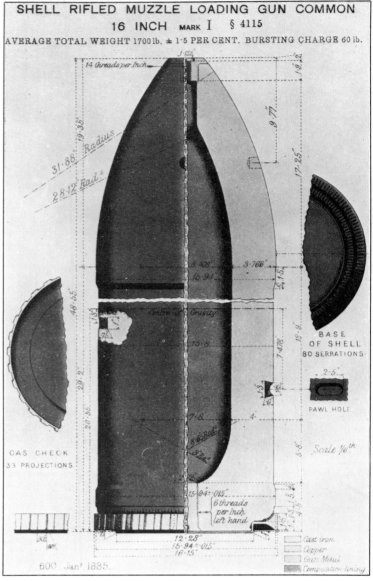

SHELL RIFLED MUZZLE LOADING GUN COMMON
16 INCH MARK I § 4115
AVERAGE TOTAL WEIGHT 1700 lb. ± 1·5 PER CENT. BURSTING CHARGE 60 lb.

BASE
OF SHELL
80 SERRATIONS.

PAWL HOLE

GAS CHECK
33 PROJECTIONS

600 Jan. 1885.

Cast iron
Copper
Gun Metal
Composition lining

(*Above*) Boxer's shrapnel shell for the 64-pounder RML. Spin was provided by the three rows of studs, while a gunpowder charge in the base of the shell expelled the musket balls when ignited by a time fuze.

(*Above right*) The 16-inch RML common shell, to be filled with 60 lb of gunpowder.

some improvements to his gun, but while he was so occupied, patents were taken out founded on his principle but without his approval, by a company called the Pneumatic Dynamite Gun Company of New York. This firm built an 8-inch weapon and engaged Lieutenant Zalinski to superintend the trials, and due to his acumen and publicity, the weapon soon became known as the 'Zalinski Dynamite Gun' and poor Mefford faded from the scene. The gun launched a projectile filled with dynamite (hence its name) and detonated by an electrical fuze of Zalinski's own design. Batteries were installed at Fort Hancock, New York, as well as in the San Francisco defences, and three were built into a special warship the USS *Vesuvius*. One was also acquired by the British Government for trials conducted in the 1890s, and it has been said that one was purchased by one of the State Governments of Australia for coast defence. But the weapons were never very successful; their maximum range was about three miles, and by the time they had been perfected conven-

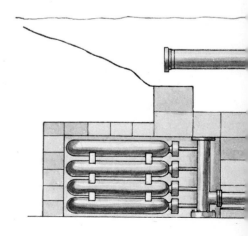

(Below) The 15-inch American Rodman gun purchased by the British Army and used to test armoured fort construction in the 1860s. (Royal Artillery Institution)

(Bottom) The Mefford-Zalinski Pneumatic Dynamite cannon installed in its emplacement for coast defence in San Francisco Harbour. The air cylinders can be seen at the front of the emplacement.

tional cannon could out-perform them with ease. However it has been suggested that most of the trouble lay in the unfortunate appellation of 'gun' to this weapon, whereas Zalinski saw it less as a gun than as a vehicle for delivering mines into local waters to supplement the existing defensive minefields; he was always very careful to refer to it as an 'aerial torpedo projector'. Be that as it may, the thing looked like a gun, was emplaced like a gun, and fired a projectile, so inevitably it was compared with a gun, to its disadvantage.

Returning to the more mundane problems, the last which awaited modernization was the matter of actually lighting the propellant charge, or, to put it in layman's language, firing the gun. Since the origin of the gun the system had been fundamentally unchanged. A vent was drilled into the gun chamber at the breech end, this was primed with gunpowder and ignited by hot iron or slow-match, thus flashing down the vent into the chamber to ignite the powder therein. Changes had been made from time to time in the method of priming and firing; in about 1765 the 'tube' made its appearance. This was a tin tube which fitted into the vent, and contained a charge of either gunpowder or quick-match – a length of cotton

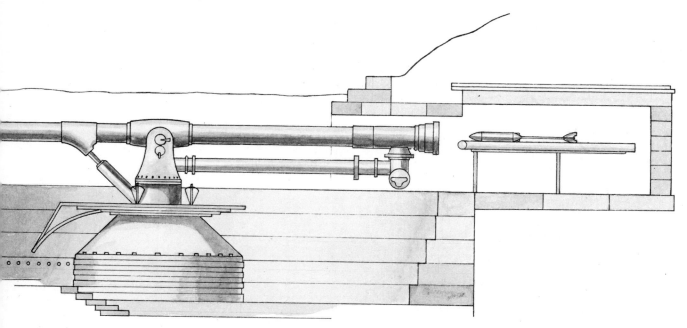

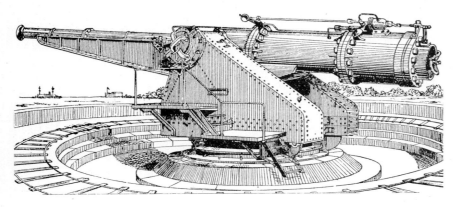

A contemporary drawing of a Zalinski dynamite gun in the defences of New York.

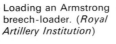

Loading an Armstrong breech-loader. (*Royal Artillery Institution*)

permeated with saltpetre – and topped with a mixture of fine gunpowder and spirits of wine. This tube was forced home into the vent and ignited in the usual way. The explosion of the charge usually ejected it after firing; these tubes increased the rapidity of serving the gun and they also eased the problem of erosion of the vent due to the back-blast from the charge, but they were somewhat weak. Many other systems of construction were tried and eventually goose-quill became the standard material in about 1780, while the filling was changed to a solid gunpowder/spirit mixture with a central hole drilled in it, which reinforced the flash and made for more certain ignition.

At about the same time the flint-lock, which had been used on muskets and pistols for many years, was proposed for use on cannon. Sir Charles Douglas suggested their use, with tubes, for the Royal Navy, but there was considerable opposition to their adoption. The metal tubes he advocated were, it was felt, a hazard to the bare feet of the sailors, and the Lords of the Admiralty

refused to adopt Douglas's idea. At his own expense he obtained a number of flint-locks and had them fitted to the guns of his own ship, and he obtained a stock of goose-quills and had tubes made to his own specification. Sir Charles was, shortly afterwards, promoted, and his ship HMS *Duke*, taken over by another officer, Captain Gardner. Gardner retained the locks and quills, and in the Battle of Gibraltar Bay in April 1782 the rapidity and efficiency of the ship's guns was very conspicuous. As a result the gun locks and quill tubes were finally approved, though they did not actually come into use until 1790.

At this time every type of gun had its own model of tube, the length being that of the gun vent, since it was thought vital that the tube be actually in contact with the propelling charge. But trials showed that this was not the case, and in about 1820 the length of tubes became standardized. The next important step came with the application of the percussion principle. Percussion caps, using fulminate compositions, had come into use for small arms in about 1815, and in 1831 the Royal Navy adopted a percussion tube, a quill containing gunpowder joined to a side quill containing fulminate. A firing hammer was fitted alongside the vent so as to strike the side tube and thus initiate the fulminate, which in turn ignited the gunpowder; the lock's position alongside the vent also made it less liable to damage from back-blast when the gun fired.

The year 1841 saw the designing of a friction tube by a Lieutenant Siemens of the Hanoverian Artillery. Here a quill was filled with gunpowder and topped with a perchlorate mixture, highly sensitive to friction. A roughened iron pin was embedded in the perchlorate so that when sharply pulled out the friction caused ignition. Siemens's

(*Below*) Methods of firing naval guns: the Navy flint-lock of 1790: a gunner's linstock with the quickmatch twisted around it; and the simple friction tube, used in land service until the end of the First World War.

(*Bottom*) Berger's cast-steel gun, one of the first attempts to use this material for ordnance construction. (*Royal Artillery Institution*)

original pattern was somewhat rough and ready but it was taken up by the British Army and, considerably improved, became the standard ignition system in the late 1850s.

The 1850s also saw considerable interest in the new science of electricity, and it was only a matter of time before somebody applied his mind to the question of firing guns with it. Benjamin Franklin, as long before as 1751, had fired a gunpowder charge with electricity as a parlour trick, and in 1853 a gun in Dover was fired by electric current from Calais, a performance which ought also to come under the heading of parlour tricks, though it does seem to be the first use of electricity to actually fire a gun. However, an electric tube was perfected in Woolwich in about 1855, largely for firing guns at proof from a distance, and these were improved until a service electric tube was issued in 1866. This, in land service, was used solely for proof firings, but the Royal Navy adopted it so as to be able to fire simultaneous broadsides.

The 1870s saw the majority of gun developmental effort poured into naval and coast defence weapons; the principal problem was still the ironclad and how to defeat it and the gun-makers were hard at work producing bigger and better guns. While Krupp and the other continental gun-

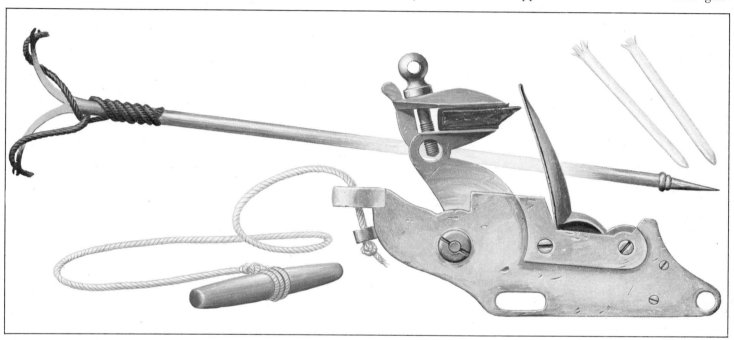

makers strove to perfect their breech-loading systems and make them compatible with powerful battering charges, in Britain they were satisfied that the RML system was the one offering most advantages all round. And there is no doubt that the heavy RMLs were the most powerful guns of their day. The 12·5-inch of 38 tons, firing an 818-lb Palliser shell by means of a 200-lb gunpowder charge could attain a velocity of 1,575 feet per second and penetrate 18 inches of iron armour at 1,000 yards range. The largest of these RML guns was a 100-ton 17·72-inch design produced by Armstrong for the Italian Government for mounting in their battleship *Duilio* and also for installation in the coast defences of La Spezia. This led to some sharp questions in Parliament, and in order to maintain parity in the Mediterranean the British Government authorized the purchase of four. These were supplied in 1878, but unfortunately the Cabinet, in its haste, had neglected to ask the Army about the problems involved, and there were no mountings suitable for the guns. It took another six years to get mountings designed and built, and even then they were largely based on Armstrong's designs for the Italian defences.

The Krupp 71-ton 16-inch gun introduced in 1879. A 440-lb charge of gunpowder was used to fire a 1,712-lb chilled shell, and on trial at Meppen it performed with remarkable accuracy, according to contemporary reports.

Systems of obturation used with screw breech mechanisms; the Elswick Cup was originated by Armstrong and used with the first 6-inch guns but proved to be too weak. The De Bange system, using a resilient pad, was universally adopted in the middle 1880s.

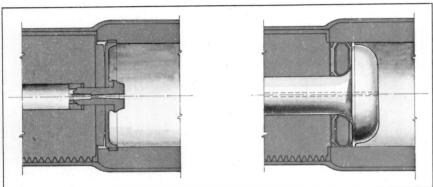

Two guns were installed at Gibraltar and two at Malta, and one of each pair is still in position. With a 450-lb charge and a 2,000-lb piercing shell, they could penetrate 21 inches of plate at 2,000 yards range. The story is told of a time when Sir Garnet Wolseley inspected the battery at Gibraltar; in the course of a demonstration firing, one charge failed to fire, and the problem arose of how to get the projectile out of the gun. Eventually it was decided that the only solution was for somebody to crawl down the bore of the loaded gun and attach a rope to the nose of the shell. A small drummer-boy was found who was willing to

The British 6-inch Mark 9 gun, one of which was bought from the Elswick Ordnance Company in 1901. A long and powerful gun with a complicated rifling system—the pitch or turn of rifling changed five times—it only remained in service for six years.

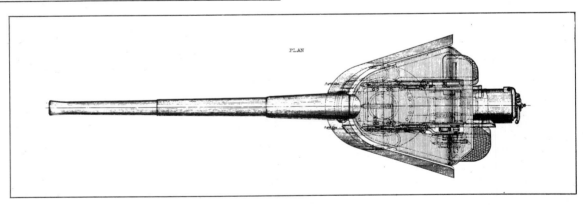

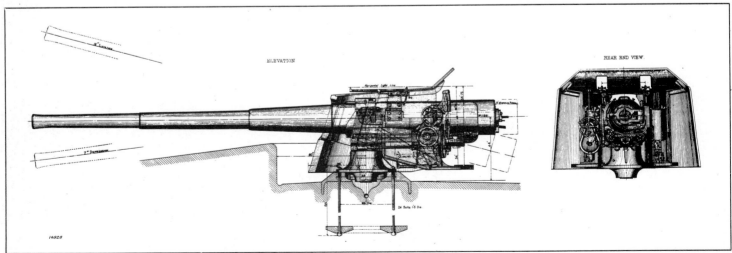

(*Above*) A photograph of the first Moncrieff disappearing carriage, with the gun 'down' in the loading position. (*Royal Artillery Institution*)

(*Right*) One way to make some use of recoil is the disappearing carriage; this experimental American 8-inch model shows the massive construction needed.

volunteer. 'You understand, do you not, that there is no danger?' asked the Commander-in-Chief. 'Not if you say so, Sir' said the boy, and disappeared head first down the bore of the cannon. The shell was successfully extracted and the drummer, according to regimental legend, was immediately promoted to corporal on his re-emergence into daylight.

But the days of the muzzle-loader were numbered. The gun-makers had been hard at work trying to produce a breech-closing system which would withstand the enormous pressures involved. Herr Krupp was by now so convinced of the superiority of his system of a sliding block that he wrote to the British Admiralty offering to supply guns of up to 150 tons weight, subsequent to a trial against a comparable Woolwich-built RML. On being asked for more details he laid down the conditions of the trial: the hire of the gun would cost £15,000; the weapon would be operated by his own men, on his own proving ground; and if his gun proved to be better, the British Government were to give him orders to the value of £2 million forthwith. This drew a dusty reaction from the Director of Artillery, who minuted that

'he cannot recommend Mr Krupp's terms to be accepted. The trials of one gun made for the purpose of competition and over which we are not to have control is inadmissible. At the rate charged for the hire of this gun (£5,000 more than the total cost of our own gun) Mr Krupp could incur any amount of expense in manufacture, particularly holding in view an order for supply costing two millions should his gun succeed. A breech-loading gun thus carefully manufactured might show none of the defects inherent in the system when manufactured for supply. All the expense of the trial, amounting to at least £50,000 would fall on us, and little would be learned from a trial carried out in this manner. We could not bind

Another view of an American disappearing carriage, this time a service model, the 14-inch gun on the Model 1907 mounting.

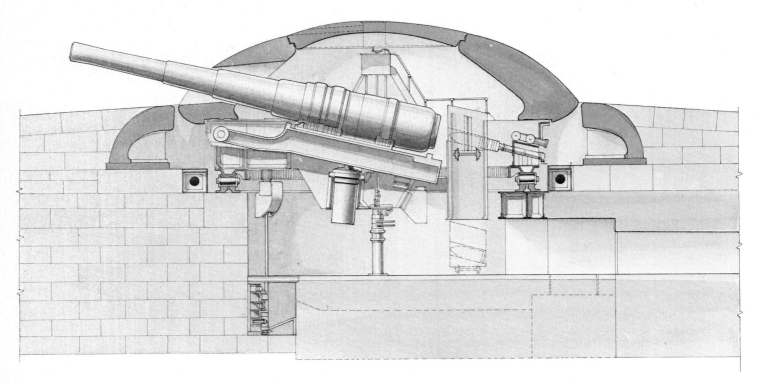

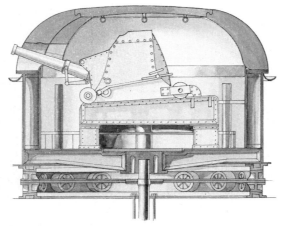

(*Above*) An example of the Gruson Cupola, often used for coast and frontier defence on the Continent. The cupola is of cast-iron armour, rotating on a roller race set in masonry. The gun is an Elswick Ordnance Company breech-loader.

(*Right*) A Fortress turret for two French 155-mm guns. Only the armoured cupola would be exposed to fire, the rest of the structure being below ground level and protected by concrete and earth.

ourselves to order guns to the amount of two millions.'

Herr Krupp's offer was politely declined.

The principal factor leading to the re-examination of breech-loading was the length of the guns. By now it was known that, other things being equal, the longer the gun, the higher the velocity, and since velocity was in demand for armour penetration, the guns were increasing in length. Moreover the new types of gunpowder demanded longer guns in which to develop their full force; firing these types of powder in short guns resulted in large portions of the charge being expelled from the muzzle behind the shot before they were completely consumed. This was not only wasteful but positively dangerous. The final blow came with the disastrous result of inadvertently double-loading a 12-inch RML gun on board HMS *Thunderer* in 1879, which resulted in the gun bursting, killing eight of the ten in the turret and killing two and injuring 35 more on the open deck. The breech-loading advocates were not slow to point out that such an accident could not have happened with a breech-loading gun, and it had to be admitted that they were right.

But already things had started to move. In November 1878 the British Director of Artillery, Sir Frederick Campbell, minuted the Secretary of State, pointing out that in view of the increased length of guns a breech-loading system appeared necessary, and that after having carefully considered the three most efficient systems of closure – Krupp's sliding block, Whitworth's continuous screw and the French interrupted screw – he had come to the conclusion that the French system was the most reliable and he had given instructions to the Royal Gun Factory to prepare a design for a gun of this type for trial. The Secretary of State concurred, and the muzzle-loading era came to a close – although the guns remained in service until 1920 in some places.

In June 1879 the inevitable Committee was formed to give an opinion as to what sort of guns ought to be built, and they were hastened in their deliberations by a sharp minute from the Admiralty which laid stress upon 'the fact that the Navy cannot wait for the best gun that laborious and prolonged investigation can produce, but dealing with information we possess and the experience acquired, the Committee are asked to give the Navy the best guns that can be made by the time the ships are ready to receive them'. The later cry of 'The Best is the Enemy of the Good' had not yet been coined, but the sentiment was there.

Second only to the guns in importance, of course, came the question of the mountings upon which they were to be placed. Field carriages showed very little change, but the requirements for mounting heavy and powerful guns in confined spaces in land defences and in ships led to much work in designing more efficient mountings. On the land front the problem was to protect the

weight was down and the gun up, the muzzle poked over the rim of the pit, nothing but the gun being visible from the sea. When the gun fired, the recoil caused it to move back, pivoting the arms to raise the counterweight, so that the gun descended into the pit for loading under cover. A friction brake held the gun down until it was required to fire again, when the brake was released and the counterweight pulled the gun up and over the pit crest once again.

Although the mounting was somewhat expensive, its installation was no more than a matter of digging a suitable pit, a much cheaper proposition than providing an expensive granite and iron-plate casemate to protect the gun, and the principle was embraced with great enthusiasm. But it then transpired that Moncrieff's design could not cope with guns more powerful than 7-inch, and the mounting of heavier weapons had to await some improvement. In the meantime they were mounted on carriages and slides of similar pattern to those used by the Royal Navy. Improvement to the disappearing carriage came by the application of hydraulics; a heavy hydraulic ram replaced the counterweight, and with this improvement,

The American 3-inch rapid-fire gun of 1903. This gun is on display in the Washington State Park of Fort Casey, one of the defences of Puget Sound.

guns and gunners from the fire of hostile ships. Ranges were short, and while in earlier times it was simply a matter of placing the gun on a truck carriage on the ramparts of the fort, the advent of better guns in the enemy fleet rendered this a hazardous proposition. The first revolutionary design came from Captain Moncrieff of the Edinburgh Militia Artillery who proposed a 'disappearing carriage' in which the gun was mounted on arms, counterweighted at their foot. The assembly was then placed on a traversing base and the whole installed in a nine-foot-deep pit. With the arms positioned so that the counter-

(*Far left*) An American 14-inch coast gun on disappearing carriage fires and begins to descend. On the left is a 16-inch on barbette mounting.

capable of being built sufficiently massive to take almost any weapon, disappearing carriages were adopted in many countries. The United States perfected the Buffington-Crozier system, which allied both counterweight and hydraulic cylinders in a design which eventually mounted guns as powerful as the 16-inch breech-loader. The St Chamond Company in France applied the principle to land fortress guns, so that they could be moved about within a fort and apply firepower to any needed point without being visible to the enemy. They were even used at sea, the Royal Navy's HMS *Temeraire* of 1876 mounting two

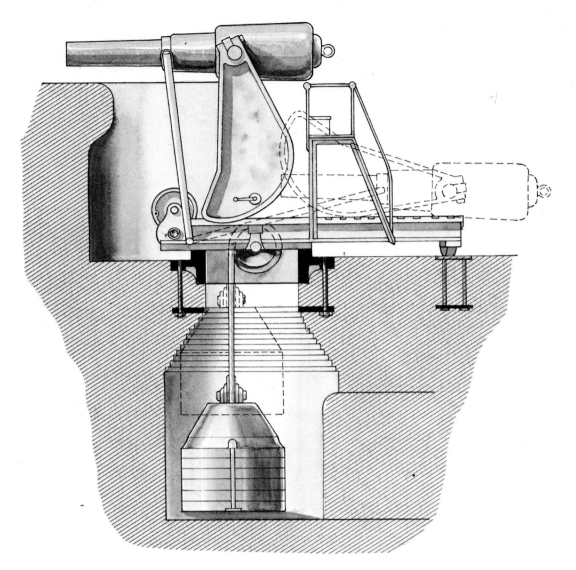

11-inch, 25-ton RML guns on disappearing mountings in barbette towers fore and aft.

On the Continent coast defences were inextric-
ably mixed up with defences on land frontiers; the
frontiers between France and Germany, Germany
and Belgium, Poland and Russia and most others
proliferated fortresses, and as guns got more
powerful the defences had to get stronger in step.
Gruson, the Ruhr ironmaster, soon carved out a
niche as an expert at producing cupolas of cast-
iron armour which were widely adopted, since
they offered a very small target and one which, in
the majority of cases, presented a curved face to
the enemy so that his projectiles would often
glance harmlessly off. In order to make the most
of this system, Krupp produced an ingenious design of mounting in which the gun muzzle was
shaped into a ball and locked into the armour of
the turret, permitting no recoil whatever. Another
popular idea was to have the cupola capable of
retraction into the ground when not in use so that
it was invisible until the enemy was actually con-
fronting them, a system later applied in the
celebrated Maginot Line.

The most difficult aspect of designing gun

mountings was dealing with the powerful recoil
of the new and heavy guns. As we have seen,
Moncrieff made the recoil do something useful, an
achievement rarely managed since, but this was an
exceptional case. Nor was it possible to adopt
Krupp's technique and simply forbid recoil to
occur by anchoring the gun in tons of cast-iron.
The hydraulic cylinder gradually emerged as the
key to the problem; a cylinder full of liquid was
attached to the gun mounting and the gun connected to the cylinder's piston rod. When the gun
recoiled the rod moved a piston head through the
cylinder. Orifices in the piston head allowed the
liquid to pass from one side of the head to the
other but offered sufficient resistance to the flow so
that the movement of recoil was damped and the
energy absorbed. When the gun came to rest, it
was returned, generally by gravity, sometimes by
springs, to the firing position. This was successful
for naval and heavy garrison weapons where the
mountings were large enough to accommodate
the necessary cylinders and springs, but it was less
successful on field pieces where the designer was
up against the old problem of how much weight
he could hang behind a horse team. It was to de-

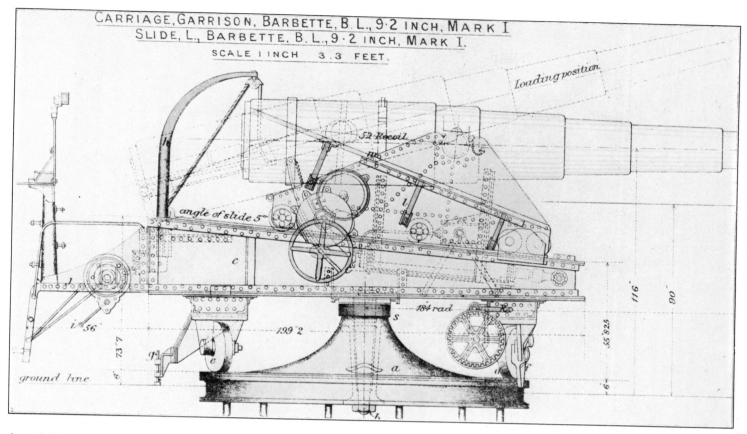

CARRIAGE, GARRISON, BARBETTE, B. L., 9·2 INCH, MARK I
SLIDE, L., BARBETTE, B. L., 9·2 INCH, MARK I.
SCALE 1 INCH 3·3 FEET.

An early British carriage for a 9·2-inch coast artillery gun; the gun and carriage recoiled up the inclined plane, controlled by a hydraulic buffer within the slide.

mand some clever designing before a hydraulic system light enough to go on to a field piece was to be invented; many said it was impossible and tried other methods. One popular idea was the 'Spring Spade' in which a hinged arm beneath the gun carriage was tipped with a spade which was lowered so as to dig into the ground. The arm was connected by a powerful spring so that as the gun carriage rolled backwards under recoil, the spade dug into the ground more deeply and the spring was stretched. While this sort of device

managed to keep the carriage in the right place, it did very little to cut down the force of recoil.

In 1897 came a surprise from France; the unveiling—but only partially—of their new field gun the 75-mm Model of 1897. By a series of what would today be called 'leaks', the world slowly got to know that the French had something revolutionary. This gun, it was said, could fire 20 aimed shots a minute, due to the fact that a perfect hydraulic recoil system was mounted on the carriage which allowed the whole affair to stay per-

One of the last attempts at a heavy RML gun mounting was this 'Armstrong Protected Barbette'. The gun could be traversed to one side, depressed, and then muzzle-loaded from beneath the cover of the pit. During this manœuvre the side of the gun presented a splendid target, which is why it was rarely used.

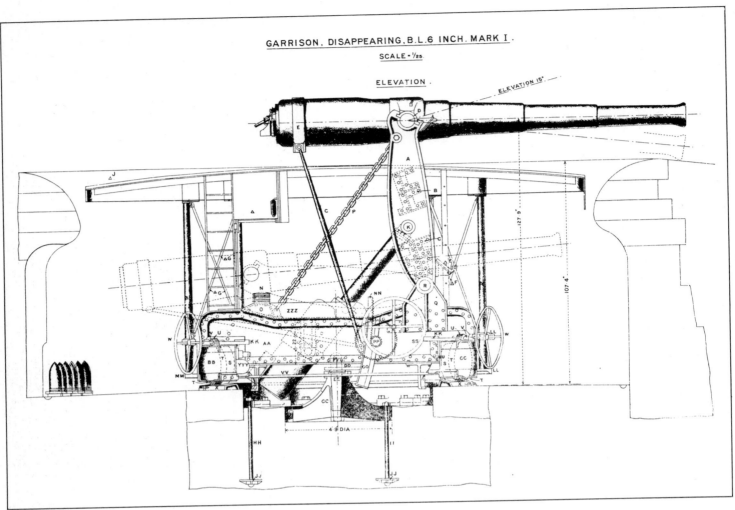

GARRISON, DISAPPEARING, B.L.6 INCH. MARK I.

SCALE = ¹/₂₅

ELEVATION.

(Above) The answer to the
protection problem was the
disappearing mounting;
this is the Armstrong-
designed hydro-pneumatic
pattern adopted in British
service in the 1880s.

(Right) A Dutch 80-mm
field gun, typical of the
early years of the breech-
loading era, before rifled
small arms and improved
shrapnel shells forced the
adoption of shields.

A British 10-inch gun on Easton and Anderson's disappearing carriage, photographed at Landguard Fort Harwich in March 1891. This design of carriage was never adopted for service. (*Royal Artillery Institution*)

fectly still while the gun recoiled. Thus the gun could be rapidly served, since the gunners could stay clustered around it; the ammunition was all in one piece, the shell being rigidly attached to a brass cartridge case carrying the charge and the ignition system, so that one movement was all that was needed to load it; and since the gunners were gathered around the gun it was provided with a bullet-proof shield to protect them. In one bound the 'French 75' made every other field gun a museum piece: the quick-firing gun had arrived.

Except for the recoil system, though, there was nothing entirely new about the individual features of the 75; it was their consolidation into one package which was revolutionary. Brass cartridge cases and fixed ammunition had been in use for several years; most navies had adopted the Hotchkiss and Nordenfelt 3-pounder and 6-pounder anti-torpedo-boat guns in the 1880s, and these used fixed ammunition with sliding block breeches, the gas seal being done by the cartridge cases. These guns, too, had used shields, as had several other naval and coast defence guns. But it was the 'long recoil' system which caused the

greatest commotion. While the French Army were careful not to reveal too much about their new weapon, observers were allowed to watch them firing from a distance and all were unanimous in their awe of the uncanny appearance when they were fired. The carriage remained still and the gun recoiled some three or four feet straight back in a 'cradle', returning almost as rapidly as it had come. The breech mechanism was a fast-acting 'Nordenfelt eccentric screw' which could be opened by the flip of a wrist to allow the empty cartridge case to be ejected and a new round to be loaded. In default of more precise information, rumour ran wild: it could fire ten miles; it had a high-explosive shell of such appalling power as to make life for the enemy impossible; the shrapnel bullets were of some secret composition which rendered them almost armour-piercing projectiles in their own right; and so forth and so on. Irrespective of what was right and what was rumour, one thing was certain—any army which didn't equip itself with a quick-firing gun very soon was going to be well behind in the race. The gun-makers reached for their pencils.

CHAPTER 5

Ready for the Fray

The French 75 proved to be a stimulus to the gun-makers, not only because of their own professional desire to go one better but because the armies of the world were hammering on their doors demanding something at least as good, if not better. Gun designs do not appear overnight, and the French had stolen a considerable march on the rest of the world. The German makers Krupp and Erhardt were the first to recover and offer quick-firing field guns of comparable design, but even so their recoil systems were inferior to the French, since they were still using springs to return the gun after recoil while the French were using a hydro-pneumatic arrangement. However, numbers of their weapons were purchased by smaller nations, while the larger preferred to wait for their own designs to mature. Among these was Britain, whose designers at Woolwich wrought exceeding slow, so slow that when the South African War broke out the British Army went forth with elderly breech-loaders on non-recoil carriages fitted with spring spades to try and bestow some measure of stability when firing. This was far from satisfactory when it was found that the Boers had outfitted themselves with some of Krupp's and Schneider's newer weapons and were making their superiority felt. In a remarkable move for that day and age, General Sir Henry Brackenbury, the Director-General of Ordnance, went to Erhardt in Dusseldorf and bought 18 batteries-worth of 15-pounder quick-firing guns complete with ammunition. These were modern long-recoil guns ranging to 6,400 yards, and with these the balance in South Africa was somewhat redressed. The national pride was rather bruised by having had to go abroad to arm the Royal Regiment, and once the war was into its closing stages the demand arose for a native product to replace the 15-pounder. It might be added that the carriage of the 15-pounder had proved to be less robust than the rest of the weapon; before being issued to the units, they were all rebuilt with standard British wheels, since the original German wheels were not strong enough to withstand much cross-country work.

As a consequence of all this, in 1901 a Committee of Royal Artillery officers met to decide on the form of the future field and horse artillery guns. After drawing up a specification they circulated it to gun-makers and sat back to await results. By this time the gun-makers had also had time to sit down and do some fundamental thinking, and they soon produced a number of designs for consideration. These were examined and tested, and the Committee decided that their requirement could best be met by amalgamating the best features of the various designs into one. As a result, the final model used the barrel developed by Armstrong, the recoil system developed by Vickers, and a carriage designed by the Royal Carriage Department of Woolwich Arsenal. This became the 18-pounder, and a smaller version for horse artillery, the 13-pounder.

The Americans, also seeking a modern quick-firing gun, took the same road as had Brackenbury and adopted the Erhardt 15-pounder as their '3-inch Model 1905'. Other nations adopted Krupp, Erhardt or Schneider designs, modified slightly to suit their national preferences, although the Russians and the Japanese, having had some practical experience during the Russo-Japanese War, elected to go their own ways and produce some designs of their own, since by now they had more definite ideas on the desirable features than anyone else.

The South African and Russo-Japanese Wars had also had their effect on tactics: in both affairs one army had gone to war with its artillery drilled in the old system of wheel-to-wheel fighting in the open, shooting over open sights at a fully visible enemy. The other had concealed their guns behind folds of ground, camouflaged their infantry positions, and used the advantages of modern high-velocity rifles with smokeless powder to pick off the exposed gunners. It was from these experiences that the first moves towards 'indirect fire' – shooting from concealed positions – began, together with such innovations as excavating gun-pits, concealing the waggons and horses, and using howitzers to search out likely positions out of view of the front-line troops.

The Russo-Japanese War had also done much to bring the high-explosive shell to the forefront. Previously the shrapnel shell had been the preferred weapon, since against troops in the open it was without equal. But the positional warfare which developed in Manchuria showed the value of high explosive in battering at fieldworks and their occupants, and more work began on perfecting a foolproof shell. One interesting variation, extensively tried in Germany, was the 'Universal Shell', all things to all men, which was essentially a high-explosive shell with a shrapnel capability. This doubtful trick was achieved by making the head of the shell a container for high explosive, so that when it was blown off to liberate the shrapnel balls it flew forward in continuance of the trajectory to act as an explosive missile. Another system was to pack the shrapnel balls in a matrix of high explosive: by opting for shrapnel fire, the fuze ignited an expelling charge in the usual way and blew the balls out, igniting the matrix to give a puff of smoke useful for observing the point of burst. If the fuze was set to give high-explosive effect, upon striking the ground the explosive was detonated and scattered the balls, and the fragments of the shell, to give a localized killing effect as well as an anti-material blast effect. In the way of compromises, these shells were never as successful in either role as a purpose-built projectile, and after the initial enthusiasm for them wore off they were gradually replaced by issues of both high explosive and shrapnel, so that the user could select

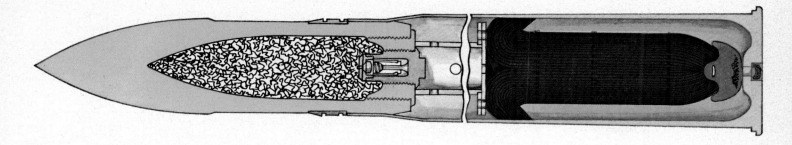

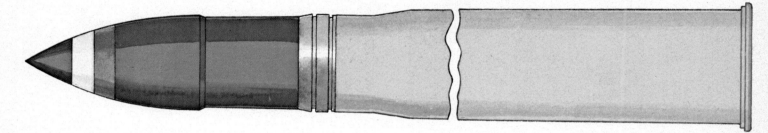

Ammunition for early breech-loaders.
(A) Star shell for the British 5-inch howitzer of 1895.
(B) Shrapnel shell for the 12-inch RML coast gun of 1871.
Both these shells worked by having the fuze fire a small
expelling charge in the base, which then blew the contents
through the weak nose.
(C) Steel common shell for the 6-pounder Hotchkiss quick-
firing coast and naval gun of 1883. The body of the shell
carried a gunpowder bursting charge and a simple fuze in the
base.

(*Right and opposite*) The Russian 6-inch gun, Model 1878, showing the method of transport, using a limber, and its arrangement in action for siege purposes.

A French 155-mm siege gun in its emplacement, showing a hydraulic brake anchoring the carriage to the ground platform, while rammers, handspikes, wash buckets and other accessories are arrayed about the embrasure.

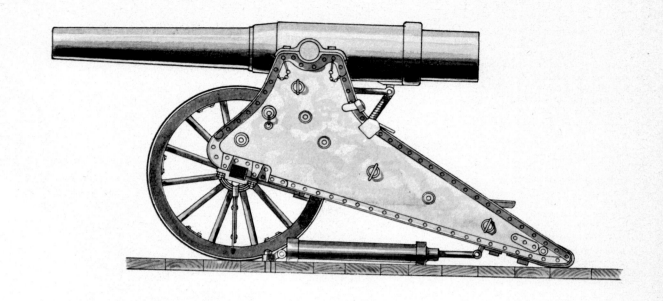

(*Right*) The French 75-mm Deport mountain gun of 1910. This also used the differential recoil system in order to save weight, the operating agency in this case being a large spring, shown in this illustration of the gun in the 'rest' position.

(*Below*) The Russian 57-mm Nordenfelt gun on coast defence mounting. Elevation was applied by the large handwheel, which operated the screw to raise or lower the breech. On recoil, the gun arm pivoted in the mounting to pull on a hydraulic buffer in the base of the mounting pedestal.

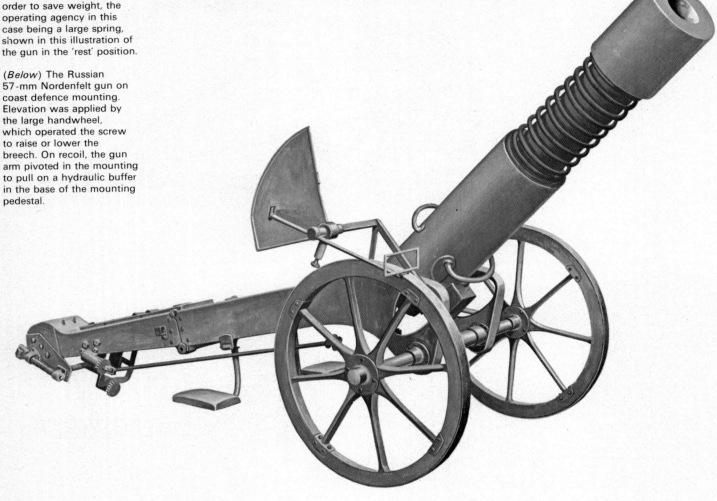

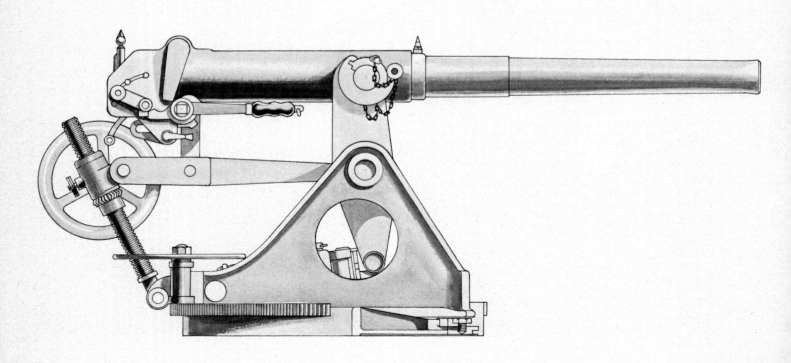

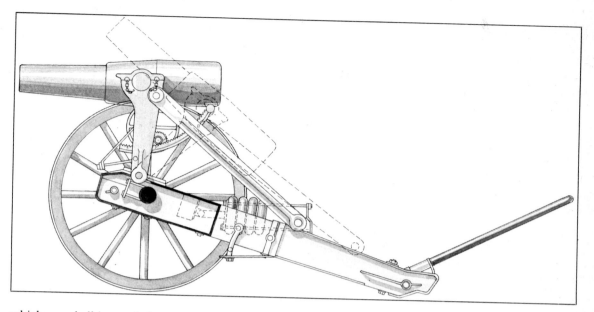

Swedish 12-cm howitzer showing an unusual system of carriage construction. The support arms are disconnected and moved down the trail so as to lower the trunnions for travelling, thus lowering the centre of gravity and making the load more stable.

(*Below*) The Spanish 15-cm howitzer, Model 1887, in section. The breech mechanism is unusual in having a hollow breech-block which formed the chamber when in place.

(*Bottom*) A Spanish 15-cm howitzer, Model 1887, mounted upon its travelling carriage and connected to its ammunition limber.

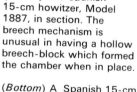

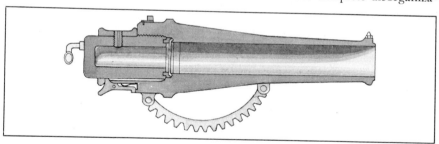

whichever shell he needed.

The high-explosive shell had also made converts in another area—that of naval gunnery. It had long been held that the only projectile worth firing from naval guns was a piercing shell in order to get through the armour protection and damage the targets behind. The Battle of Shimonoseki, or Tsushima, where the Japanese fleet conclusively defeated the Russian, brought some interesting points to light. It appeared that the Japanese had relied very much on high-explosive 'common' shells rather than piercing projectiles, and these had shredded the upper works of the Russian ships, causing innumerable casualties and starting fires sufficient to cause the complete disorganiza-tion of the ship as a fighting machine without having to go to the extent of piercing the armour and sinking it.

By the time the new century was a year or two old, gun design had settled down into a few well-tried and trusted systems, and apart from a few minor aberrants, was set in the mould in which it has more or less remained. The guns were built according to one of three standard methods: they were either built-up, wire-wound or single-tube designs. The built-up gun was nothing more than Armstrong's original system brought up to date. The material was now invariably steel, and the gun was begun with the actual rifled barrel tube; this was then reinforced by shrinking successive hoops on to surround the chamber and the adjacent part of the barrel to give strength where it was most needed. The improvement on Armstrong lay only in the choice of material and in minor matters of the contours of the hoops and how they were interlocked. The wire-wound gun, on the other hand, began in the same way with a steel rifled tube, but this was reinforced by wrapping miles and miles of ribbon-like steel wire round it under tension. The effect was to compress

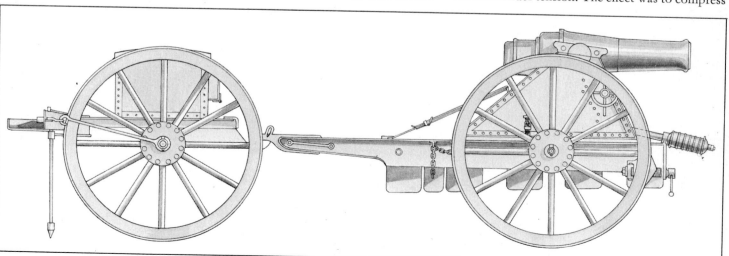

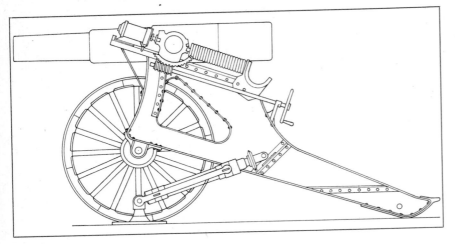

The carriage of the American 7-inch siege howitzer Model 1890. Recoil was damped by a bank of Belleville springs behind the trunnions, while the violence of the gun's return was damped by a small hydraulic buffer in front. There was also a ground buffer connecting the carriage to its platform.

Americans never took to them. Droop can, in fact, be compensated for in the design of the weapon's sights, and is no drawback to accuracy; indeed, the early issues of the British wire-wound 18-pounder were greeted by storms of abuse — 'damned crooked guns' was one of the milder phrases applied — when they were received by the gunners, but after using them, and after being assured by the ballisticians that there was a momentary straightening of the barrel at the instant of firing, they were accepted and turned out to be as accurate as anyone could wish.

The second drawback to wire-wound guns was that when the gun wore out and required a new barrel it was necessary to remove the hoops and then unwind the miles and miles of wire; with a built-up gun this was unnecessary. As a result of this it became the normal practice to make the basic barrel a smoothbore, wrap it and build the gun, and finally slide into the bore a rifled 'inner liner', forcing it home by hydraulic rams. Thus, when the time came for relining, the inner liner was forced out again by hydraulics and a new one replaced, without disturbing the rest of the gun.

The single-tube gun, as its name implied was no more than a block of steel which was turned to the necessary shape on the outside and bored

the barrel and resist the explosive forces within by the effect of the many individual hoops of wire formed round it. After the wire, hoops of steel, similar to those of a built-up gun, were shrunk on, but the principal resistance to bursting was that of the wire winding. While a very reliable system, it suffered from two defects. In the first place the wire winding did nothing to support the gun's length, and thus wire-wound guns were notorious for their 'droop' — one of the reasons that the

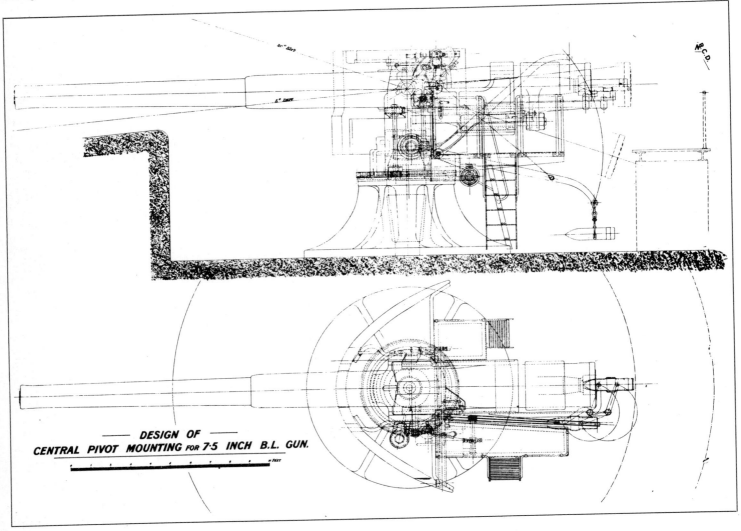

DESIGN OF
CENTRAL PIVOT MOUNTING FOR 7·5 INCH B.L. GUN.

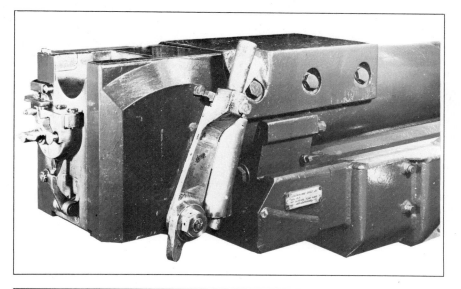

the outer layers of metal placed the inner layers under compression and thus strengthened the gun much as the shrinking on of a hoop would have done. This technique could be used with a single-tube gun or with the basic barrel of a built-up gun, and in the latter case reduced the number of hoops necessary to give the required strength.

Between the gun and its carriage came the recoil system, and here the designers had much scope for individuality. Recoil was damped out by a hydraulic buffer, which consisted of somehow drawing a piston head through an oil-filled cylinder. One could attach the cylinder to the gun and pull it across the piston head, or attach the piston to the gun and drag it through the cylinder; it all amounted to the same thing, though pulling the cylinder added to the recoiling mass and helped keep the gun stable. As the piston moved through the oil a port or valve allowed oil to pass

(Top) The vertical sliding block breech in typical form; a percussion firing mechanism is mounted in the block.

(Above) The sliding block breech opened; the breech operating lever is checked by a spring buffer.

(Right) An early Krupp design of 65-mm anti-aircraft gun, about 1910. The tip of the trail is anchored and the wheels swung through 90 degrees to allow rapid swinging of the gun to follow an aircraft in flight.

(Left) The Indian government demanded a more powerful weapon than the 6-inch for their coast defences, and adopted this 7·5-inch in its place.

and rifled on the inside. An alternative to this was to force a mandrel through a red-hot billet of metal, compressing the inner fibres and adding some strength to the gun; but these systems were only used for relatively small weapons firing less powerful cartridges and it was never widely adopted.

At this time, early in the century, a fourth system of gun construction was being explored, though it was not until the 1930s that it was sufficiently perfected to allow of its use in service. This was the 'cold-working', 'radial-expansion' or, to use the French term which is now common 'auto-frettage' system. In this system the gun tube was forged slightly below its final diameter and then subjected to an interior hydraulic pressure which expanded the bore by about 6 per cent and the exterior by about 1 per cent. When the pressure was removed the outer layers of metal tended to return nearly to their original dimension due to the elasticity of the steel, whereas the inner layers, expanded much more, tended to retain their enlarged diameter. The result was that

from one side to the other, and as the recoil stroke progressed so this port gradually closed and recoil came to an end due to the incompressibility of the oil. Once the gun was brought to rest it had then to be sent back, ready to fire again. Here two systems were available: one either used compressed air (or some inert gas such as nitrogen) or banks of springs, and accordingly the system became a 'Hydro-Pneumatic' or a 'Hydro-Spring' system. The hydro-spring system was the most simple, since the springs could often be wrapped around the hydraulic buffer to make a very compact unit. Using hydraulics was usually more complicated, which is why few people tried it in the early days. The basic requirement here is to have a second cylinder and piston arrangement which, during the recoil stroke, are moved so that

An example of the interrupted-screw breech mechanism for use with bagged charges. A firing lock, a miniature sliding block breech mechanism, is attached to the rear of the breech-block carrier.

The screw breech opened, showing the stepped threads of block and breech aperture, and the dark obturating pad in front of the breech screw.

air is compressed behind the piston head. Obviously, a fair amount of pressure is going to be needed to return several hundred pounds of gun, so the air is usually highly compressed to begin with, 600 lb per square inch being a good average figure for a field gun.

These are the basics; there are many more refinements to be added before one arrives at a working recoil system. For example, as the gun fires the energy delivered to the system leads to the heating and expansion of the oil, and some system has to be incorporated to compensate for this; whether the gun is returned by springs or air there must be some sort of fluid damper arranged

so that it comes to rest gently and not with such a bang as to overset the gun on to its muzzle. Then there is the problem of howitzers and high-angle guns; if the gun is to be stable at point-blank the recoil stroke needs to be long, to absorb much of the energy. But when the gun is elevated, the recoil stroke cannot be too long or the breech may strike the ground, and therefore another compensating device has to be built in to reduce the recoil stroke as the gun elevates. All these, and more, problems were met in the 1900s and each was, in turn, conquered. Some peculiar things happened in the process. There was, for example, the 'Differential Recoil' system. This was adopted in some mountain guns where the lightest possible equipment was demanded, because it dispensed entirely with half the system—the hydraulic buffer. The recoil system consisted solely of the recuperator springs; before firing the gun was pulled back manually to the fully recoiled position and locked there by a catch. It was then loaded; to fire, the catch was released allowing the gun to run forward. A fraction of a second before it reached the end of its run, the firing mechanism was released and the gun fired. The recoil force now had to arrest the forward-moving mass, reverse its motion, and drive it back, and this reversal caused much of the recoil force to be absorbed so that the springs, doubling now as recoil springs, were sufficient to deal with what force remained. As the gun recoiled, so it was caught and held against the springs ready for the next shot.

The whole thing sounds ridiculous, but is in fact based on very sound principles—after all, most sub-machine-guns work on a similar principle. The principal drawbacks were the question of arranging it so that various strengths of charge could be fired, since each would have differing recoil characteristics, and the question of what would happen if the cartridge misfired. The gun then went over on its nose due to the force of the forward-moving mass, and if the misfire turned out to be a hangfire and the cartridge went off, the result could be highly embarrassing. Nevertheless, the theory gets taken out and aired from time to time: it popped up in Germany in 1943, applied to an anti-aircraft gun, and it is at present being experimented with in the United States under a new name, 'Soft Recoil'. According to the newspaper reports it is now all done by electronics, which serves it right.

The gun carriage had also become more or less standardized. Bearing in mind that the horse was still the universal motive power, lightness was the aim, and the usual field gun was a simple construction of axle-tree, wheels and a pole trail. Traverse was provided either by placing the gun and recoil system on a 'top carriage' and pivoting it on the lower carriage, or by 'cross-axle traverse' in which the axle was formed into a screw-thread and the whole carriage was moved across it by a handwheel attached to the carriage and working in to the thread. This gave a relatively small

A French 120-mm siege gun, also attached to its ground platform by an hydraulic buffer. Chocks were provided to give an additional control of recoil, and the carriage has an extra set of trunnion mountings for the barrel to be displaced into for travelling.

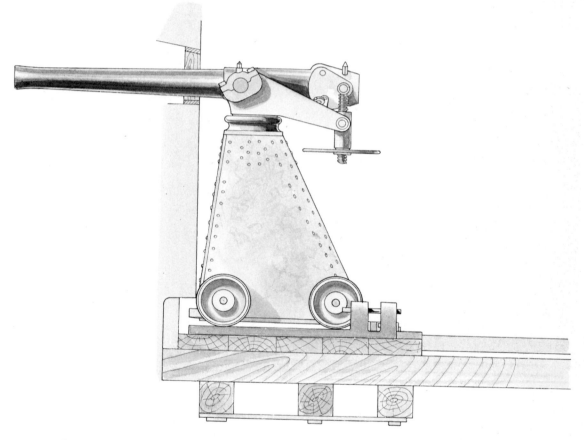

Another mounting for the Russian 57-mm, this time for the defence of fortress caponiers. The wheeled carriage allowed the weapon to be moved about the fort as necessary. Note that, in contrast to the coast defence mounting, there is no recoil system.

amount of traverse, but since the guns were light they could be easily moved by one man heaving the end of the trail across. The disadvantage of this class of carriage was that since the pole trail was central beneath the gun, it followed that after a relatively small amount of elevation of the gun, the breech end would be in danger of striking the trail, and this consideration restricted the amount of elevation available. The average amount was 16 degrees, which gave a range in the region of 6,000 yards with contemporary ammunition. In the 1900s this was thought to be sufficient, since most shooting was carried out at less than this range, the guns being commanded and the fall of shot observed from the battery area. For howitzers, firing at high angles, it was customary to make the trail in the form of a parallel-sided box so that the breech could recoil through, but this carried its own drawback in that the gun could only be traversed a small amount before the breech struck

A 15-cm Skoda gun of the Portuguese Army, dating from the 1920s.

A British 6-inch 26-cwt howitzer, introduced during the First World War and which remained in service until the middle of the Second.

the sides of the box. An Italian designer came up with the best answer: he split the trail into two separate legs which could be locked together into a single unit for towing the gun, and then opened and spread to form a wide angle when firing. With this system there was little or no restriction on traverse until it reached such an extreme that the gun was virtually firing across the trail, at which point it tended to upset the whole structure; but this was only reached at very wide angles, and provided the traverse was held to about 50 degrees there was little danger.

The field gun was generally completed by the provision of a shield of shrapnel-proof armour plate, about a quarter of an inch thick, mounted ahead of the axle. The only problem now remaining was that of aiming the gun and hitting the target, and here there was a divergence of opinion. The basic principles were in no doubt: to deliver the shell to a certain range the barrel had to be elevated by some specific amount; it also had to be pointed in the right direction; and since the rifling of the barrel caused the projectile to drift off to one side, this had to be taken into account somehow. Given these three facts, every maker had his own solution: drift, for example, could be corrected by setting the rear sight at an angle, so that the more the sight was elevated the more divergence there was between the line of sight and the line of the piece; or one side of the carriage could be made slightly higher than the other so that the gun was always offset in the opposite direction to the drift. But then, a similar effect could arise if the gun's wheels were not resting on level ground, and thus some form of compensation had to be introduced for this, while still allowing for the drift.

Elevation was generally applied by mounting the rear sight on an arc graduated either in degrees of elevation or in actual range; the gun-layer moved the arc until the graduation was opposite an index mark, which depressed the line of sight. He then elevated the gun until the line of sight came back on to the target, by which movement the gun had been elevated by the desired amount. Another solution, favoured by the French, was the 'independent line of sight' in which one layer pointed his sights at the target to give direction, while on the other side of the gun a second layer applied elevation by working from

Typical of the smaller coast defence gun is this 4·7-inch model, for protection of ports from attack by fast light destroyers.

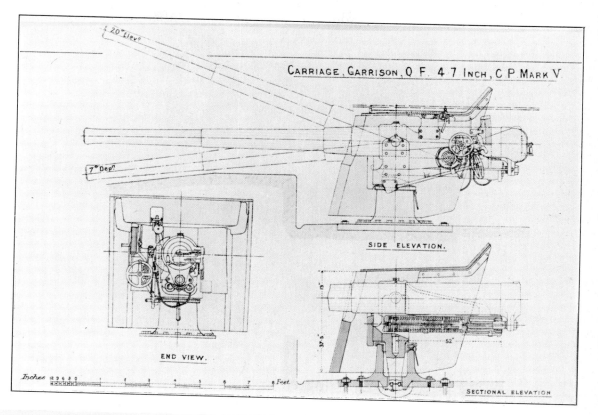

The 9·2-inch devoid of its mounting, and (*below*), installed in a coast battery near Plymouth.

a graduated spirit level.

Pointing the gun at the target was done by directly lining up a rear and front sight, like those of an ordinary rifle, or by a telescope. These sufficed until indirect shooting became normal practice, after which something more involved was needed. After some elementary ideas had been tried, this settled into what was known as the 'panoramic' or 'dial' sight, a form of periscope with a rotatable head graduated in degrees. When the gun was brought into action some prominent object in the area was selected as an 'aiming point' –for example a windmill or church steeple. With this point and the gun located on the map, a few minutes' work sufficed to work out the angle between the aiming point, the gun and the target. This angle was set on the sight head and the whole gun shifted round until the sight pointed at the aiming point, whereupon the barrel was pointed at the target. So long as the gun was always relaid by bringing the sight back on to the aiming point, the gun remained pointing at the target.

Coast defence guns were the same as field guns except for their greater size, using the same methods of construction and similar recoil systems, though these could be smaller in size and recoil stroke since the whole weapon was firmly anchored in concrete. The mountings were relatively simple; the day of the disappearing carriage was passing, since due to their odd geometry they could not elevate to high angles for long-range firing. Light guns sat on top of simple pedestals, while the heavier equipments adopted the 'barbette' mounting, a large steel drum on which the gun rotated on a roller race, protected by a shield and a concrete emplacement. Aiming was usually direct and a form of telescope known as the 'automatic sight' was evolved. This relied on the fact that a coast gun was always at a fixed height above sea level, and thus this height could be used as the base of a triangle, one other side of which was the horizontal range. Hence the sight became a simple form of rangefinder: by mounting the telescope on a suitably shaped cam, laying the telescope on to the bow wave of the target vessel automatically gave the correct elevation to the gun. A compensation had to be made for the rise and fall of the tide, and for the curvature of the earth, but that was a relatively simple matter. For larger weapons firing at ranges at which the gunlayer would be hard put to see the target, observation was done by powerful optical range- and position-finders some distance away from the guns, and their observations converted into gun information by a primitive form of computer. These data were then sent to the guns by telephone and the range and bearing applied against scales in the mounting, the guns firing without their detachments ever seeing their targets.

As the century drew on, a new problem presented itself – the airship or aeroplane. The first 'Quick-firing Balloon Guns' as they were called, were publicly exhibited at Frankfurt in 1909 by Krupp and Erhardt. The problems appeared enormous in those days: almost every aspect of anti-aircraft fire was open to debate, from the type of projectile to be used – shrapnel, high explosive, or some form of incendiary to ignite the hydrogen in the airship – to methods of sighting, fire control, the development of fuzes sufficiently sensitive to

be operated by contact with balloon fabric, and even the problem of what was going to happen to the shells which went up and failed to hit the target. A book published in 1910 was somewhat offhand about this prospect:

'It has been objected to balloon guns in general that our own troops will be endangered by the shells falling on their own heads. This objection is, however, unsound; even if a balloon is attacked by a rival dirigible or aeroplane, it has to be destroyed by projectiles of some sort. And it matters little to the soldier below whether a shell which falls on his head from 5,000 feet weighs one pound or twelve. Moreover the Krupp 12-pounder ranges some eight miles at 45 degrees elevation, so that at any rate the troops in the vicinity will not suffer. Finally, since the object in view is to bring down some tons of balloon or some hundredweights of aeroplane from the sky, the incidental fall of a few 12-pound shells would appear to be a minor matter.'

Put like that, it all sounded very convincing, but when it actually began to happen around London in 1915, it wasn't dismissed quite so lightly.

Two lines of approach to the aerial problem were examined: the Krupp answer was to develop a 3-inch high-velocity gun firing a high-explosive shell with a 'smoke-trail' fuze so that the flight of the shell could be followed and corrected and the burning composition would ignite the balloon if it struck without detonating. Erhardt preferred to use a lighter, 6-pounder, gun and mount it on a

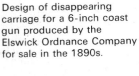

Design of disappearing carriage for a 6-inch coast gun produced by the Elswick Ordnance Company for sale in the 1890s.

The Gatling gun. When
the early mechanical
machine-guns were intro-
duced they were invariably
employed as field artillery,
but their totally distinct
function was later recog-
nized and they were
removed from their field
carriages.

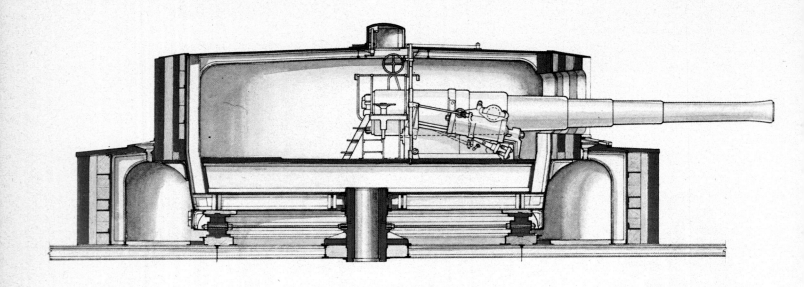

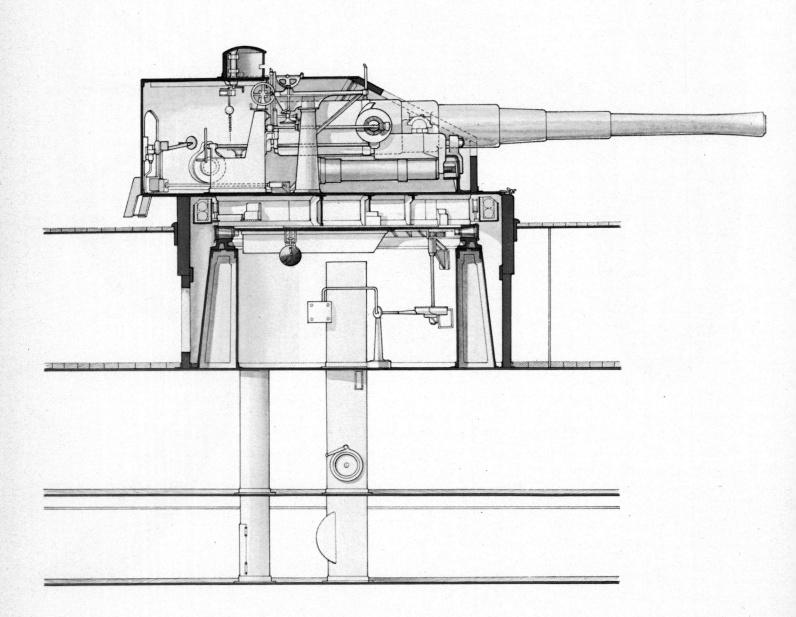

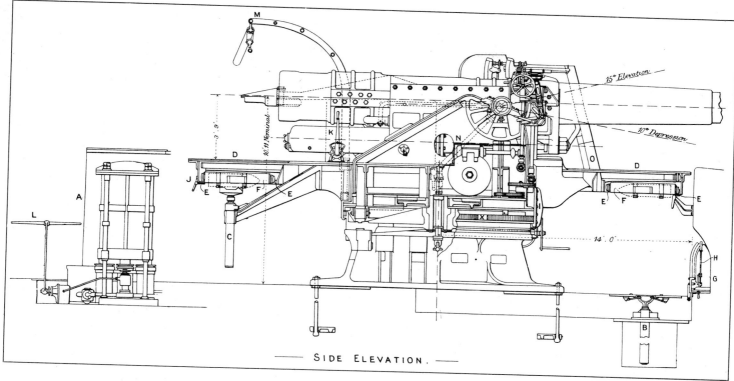

— SIDE ELEVATION. —

(*Above*) An installation drawing of a British 9·2-inch coast gun. A hydraulic lift (B) raised the shell to a rail track beneath the shell-pit shield (F). From here it was run round to the second hydraulic lift (C) which raised it to the breech. An accumulator (A) at the rear of the pit was pumped up by members of the detachment and provided hydraulic power.

(*Opposite*) Some early naval turret designs. The top view is of an Elswick Ordnance Company design for a single 9·2-inch breech-loader, while the side view shows the two-gun turret for mounting 10-inch guns in HMS *Thunderer*. Both were produced in about 1890, but the Elswick design was power operated while the naval design was hand worked.

motor-car chassis, the object being to chase the intrepid aeronaut and destroy him by rapid fire at short range. The French also inclined to this theory of mobile pursuit and began experimenting with their ubiquitous 75-mm Model 1897 mounted on a De Dion chassis.

This use of motor chassis for mounting guns can be adduced as an early forerunner of the self-propelled gun, but apart from this the motor vehicle was beginning to make its presence felt in the world and it was no surprise when people began to suggest that it might be a more efficient way of taking guns into action than pulling them there behind a team of horses. The traditionalists didn't like the idea of losing their horses—they still haven't got over it—but more and more trials were made. In Britain, one small straw in the wind was a demonstration mounted by a Territorial battery who borrowed a number of Sheffield-Simplex tourers, hooked their 15-pounders on behind, packed the gunners into the open cars and made a spectacular 120-mile drive across Yorkshire during the summer manœuvres of 1914. It was all very convincing, but even the most fervent supporters had to admit that the motor car was at a disadvantage once it left the highroad. The Austro-Hungarian Army took two 24-cm howitzers into their 1912 manœuvres towed by motor trucks, while the Italians had announced their 'FIAT Motor Limber' towing a 15-cm howitzer in 1911. The French, too, were looking at the motor vehicle for at that time their motor industry was probably more advanced than anyone else's, and in the 1912 war games they produced a battery of 12-cm siege howitzers, one of 21-cm siege howitzers and a single 24-cm howitzer, all towed by motor trucks. But it all came back to the

questions of cross-country performance and the availability of skilled mechanics to look after these modern contraptions. The supply of fuel and lubricants and spare parts was also a daunting problem; you could generally find something to feed a horse on, but finding petrol could be difficult in those days. Most of all it broke on the rock of skilled manpower; at that time almost every soldier had some knowledge of horses before he enlisted, even if it was only to distinguish which end bit and which end kicked, so that training him to look after horses was not too difficult. However, men who understood the motor vehicle rarely found their way into the Army, and until either recruits got brighter or motor vehicles more simple, most armies were content to stay with their horses.

It is difficult for the reader in the 1970s, when soldiers are seen on all sides manipulating such things as computers, missiles, lasers and other highly technical devices, to appreciate the fear in which anything remotely difficult was held by the military authorities in the 1900s; it was an article of faith that the average soldier was so stupid that he could barely be trusted to hold a hammer by the correct end unless supervised, and the adoption of many useful devices was resisted by the authorities on the grounds that it would be impossible to instruct soldiers how to use or care for them. It took the First World War to demonstrate the fallacy of this viewpoint.

Among navies, on the other hand, this attitude had long since been swept away and complicated machinery was the order of the day; space and manpower were both at a premium on warships, and after some initial misgivings it was appreciated that machinery could be used to save on both.

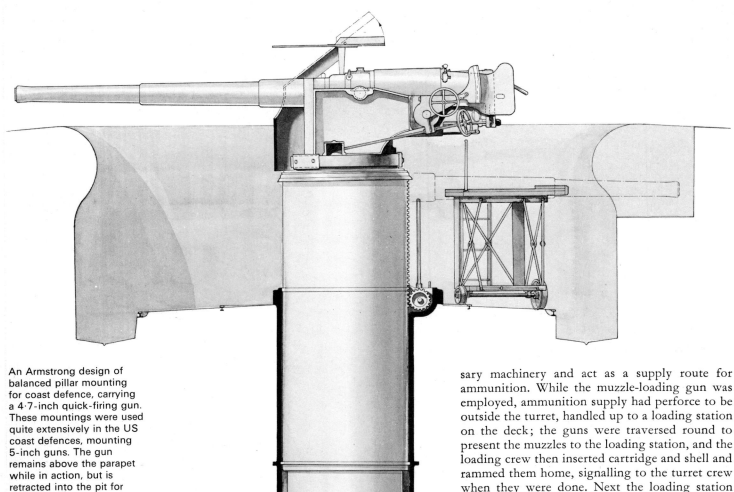

An Armstrong design of
balanced pillar mounting
for coast defence, carrying
a 4·7-inch quick-firing gun.
These mountings were used
quite extensively in the US
coast defences, mounting
5-inch guns. The gun
remains above the parapet
while in action, but is
retracted into the pit for
concealment at other times.

The cumbersome recoil slide and carriage gave
way to 'broadside mountings', compact mountings
in which the gun recoiled on a short inclined plane,
controlled by hydraulic buffers and returned by
springs, and these in turn gave way to pedestal or
central pivot mountings in which the gun was held
in a 'cradle' and allowed a short axial recoil, again
hydraulically controlled. So by these successive
improvements the mountings took up less space,
required less men to operate them and, into the
bargain, were capable of carrying more powerful
guns.

It was in the turret installations that machinery
really came into its own; Cole's original turrets
were little more than iron boxes revolving about
a pivot in the deck, but this simplicity was soon
left behind. The great weight and power of two-
and three-gun turrets demanded more secure
forms of mounting, leading to roller races and
heavier pivots, and then came the idea of making
the turret an almost self-sufficient unit by sinking
it into the structure of the ship for stability and
using the trunk of the turret to carry all the neces-

sary machinery and act as a supply route for
ammunition. While the muzzle-loading gun was
employed, ammunition supply had perforce to be
outside the turret, handled up to a loading station
on the deck; the guns were traversed round to
present the muzzles to the loading station, and the
loading crew then inserted cartridge and shell and
rammed them home, signalling to the turret crew
when they were done. Next the loading station
was given more protection by placing it beneath
the deck under an armoured shroud, the gun
muzzles being depressed below deck level for
loading. But with the adoption of breech-loading it
became necessary to get the ammunition into the
turret, and this brought the turret to the zenith of
its design. The first stage was simply to install an
open hoist in the body of the turret which allowed
the ammunition to be hauled up from the magazine, handed into the turret and then rammed into
the guns. While this was reasonably efficient it
carried a considerable hazard in that should the
combustible gases left in the gun breech after
firing flash back when the breech was opened, as
sometimes happened, the flame could ignite the
next cartridge awaiting loading and the resulting
flash could pass down the open shaft into the
magazine. The point was not appreciated until it
actually happened, the United States Navy in particular suffering some very serious accidents of
this type; on 13 April 1904 the USS *Missouri* was
at firing practice, and while loading one of the
12-inch guns in the turret, the propelling charge
exploded, igniting four more cartridges in the
hoist, a total of almost half a ton of gunpowder.
Thirty-two officers and men were killed.

As a result the next step was to introduce automatic flame-proof doors in the structure to seal
the turret from the ammunition handling room
and this from the magazine. This answer was not

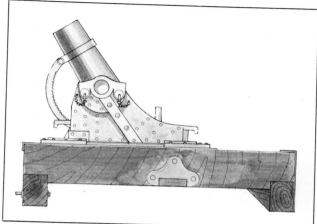

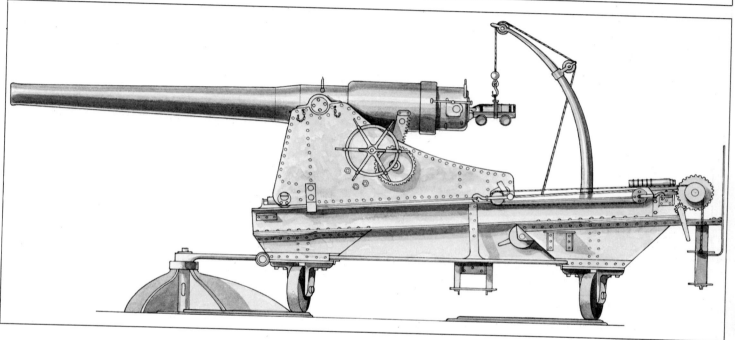

(*Top, left and right*) An
Italian 15-cm mortar of the
siege train, showing its
method of transportation
and the system of placing it
in action on a 'mortar bed'.
The weapon was a breech-
loader, using a sliding
block breech based on
Krupp patents.

(*Above*) An Austro-
Hungarian coast defence
gun of about 1885, typical
of its type and time.

entirely satisfactory, and the next step was to split
the hoisting movement into two stages, with one
set of hoists lifting ammunition from the bottom
of the turret to an intermediate handling stage,
where it was transferred to a second hoist to complete the trip up to the gun. By fitting the hoist
doors with interlocking mechanism it was impossible to have both sets of doors open at once,
which gave a more positive insurance against
back-flash accidents.

More and more machinery made its appearance:
in order to get rid of the combustible gases which
caused back-flash, guns were now fitted with air-
blast pipes which blew a high-pressure stream of
air through the bore before the breech was
opened; the effect of this can often be seen on
films of naval guns firing, when, after the blast of
discharge, there is a short and silent puff of smoke
from the muzzle as the air-blast clears the bore.
Another danger lay in the possibility of the bagged
charge leaving behind some smouldering residue
which, fanned into flame by either the air-blast
or the draught caused by opening the breech,

could well ignite the new cartridge before the
breech was closed behind it. To deal with this
hazard, wash-out pipes forced a spray of water
into the gun chamber before the breech was
opened. As a final insurance the whole interior of
the turret was frequently sealed and supplied by
pumps with pressurized air, so that when the
breeches were opened any poisonous fumes from
the exploded charge not cleared by the air-blast
would not be able to enter the turret due to the
pressure differential.

As the machinery increased so the space grew
less, and to save manpower more work was taken
over by machines: the breeches of the guns were
opened and closed by hydraulic motors; ramming
of the shells was done by chain rammers driven by
hydraulic motors; elevation of the guns and
revolution of the turret was done by hydraulic and
later electric power. Above all, all this mechanical
apparatus was interlocked so that it was impossible, for example, to ram unless the breech was
open, or to operate the wash-out if the gun had
not been fired. A refinement which might not be

thought of by the layman was the provision of cam-operated safety gear connected to the hydraulic training and elevation mechanisms so that as the turret revolved the guns were automatically elevated and depressed so that at no time were they in danger of striking or shooting at any part of the ship's superstructure; as an additional insurance, this mechanism was interlocked with the electric firing gear of the guns.

With all this sort of machinery, plus two, three or even four guns of up to 14-inch calibre or more, it is not surprising that the turret structure now weighed up to 1,000 tons and became a highly specialized manufacturing problem. Commercial companies such as Krupp, Vickers-Armstrong, Schneider and Ansaldo became recognized experts in the subject and devoted large workshops to turret construction; for in order to be sure everything worked properly, the turret was actually built up in a special pit within the factory in the first instance, tested and adjusted, and then dismantled and installed in the intended ship. It is hardly surprising that a set of turrets could take two years to build and fit, as much as the rest of the vessel, and when sudden naval expansion was called for, turret-building often became the critical factor in the construction programme.

Having got these guns into a ship, two problems remained: first to control them so that the projectile actually arrived at the desired spot; and second to make sure that when the projectiles did arrive they did the required amount of damage. The days of allowing each gun captain to pick his point of aim and let fly as he pleased once action had been joined were long gone; nowadays fleets were liable to come to blows at ranges where the men manning broadside and turret guns would scarcely be able to see their opponents. And with targets at such a range, it required optical assistance to locate them, determine the range and take some sort of aim. Thus there came about the gradual centralization of fire control into one place in the ship, equipped with rangefinders and other spotting aids, and normally located as high on the vessel's superstructure as could be arranged. This armoured cell transmitted data to the guns and controlled their firing, so that the guns fired together instead of at the whim of the individual gun captain. This gave the advantage that the sudden arrival of a broadside was a more devastating affair than the piecemeal arrival of the same number of shells, and, more important, it ensured that the spotters knew what was going on. If the guns were allowed to fire as they chose, it would have been impossible to determine which gun fired which shell on what data, and thus making an intelligent assessment and correction from observing the fall of shot would have been impossible.

Naval gunnery thus divided itself into two portions: firstly the determination of the range and bearing to the target and the conversion of this information into figures which could be passed to the guns; and secondly the ability of the gun-layers to set this data on their guns and fire them. If the first part were done properly, then the second should be infallible, provided the gun crews were properly trained, and, conversely, if the crews were up to the mark any error was likely to come from the former part of the business. In consequence much thought went into the perfection of the data from the moment of the basic determination of the range and bearing to the moment of its arrival at the gun. The problems of drift correction, to counter the sideways movement of the

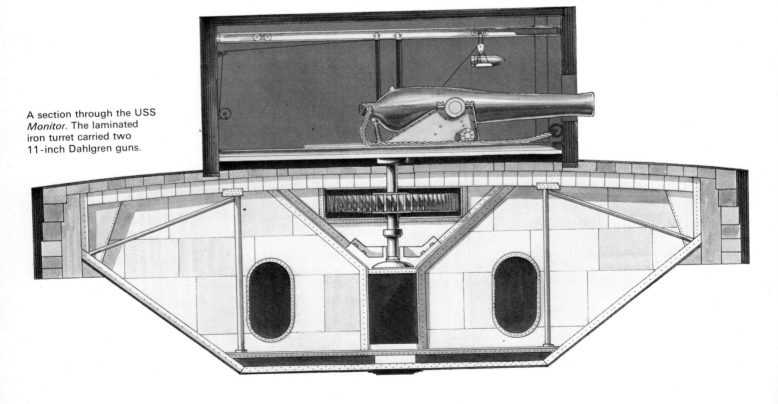

A section through the USS *Monitor*. The laminated iron turret carried two 11-inch Dahlgren guns.

The British 6-inch 30-cwt howitzer of 1895. While nominally a field army weapon, it could also be fitted to a special ground platform to become a fixed siege piece.

A British 9·2-inch gun on 'Vavasseur' mounting; Vavasseur's principal feature was the use of inclined planes up which the gun recoiled, coupled with hydraulic buffers. This is a naval barbette version, with a light shield and a working platform at the rear for the gun's crew.

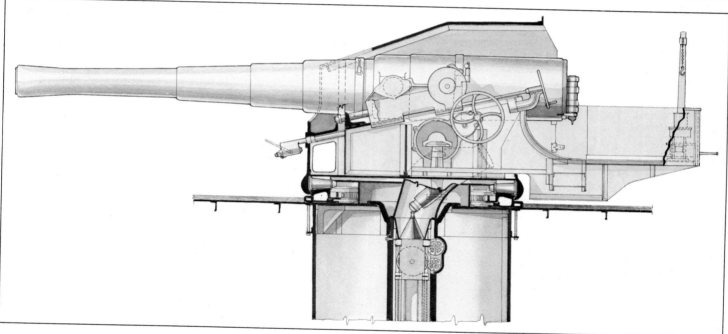

shell due to spin; corrections needed to compensate for the action of the wind on the projectile; adjustments to cater for the fact that the gun wore away gradually and thus achieved a lesser velocity with each shot; adjustments to compensate for the change in performance of the propellant as the temperature varied; all these were gradually taken into account and, as they were eliminated, other phenomena made their appearance. For example, the roller path supporting the gun mounting might not be perfectly level with the deck, and at certain angles of training the gun barrel could be at a higher or lower elevation than that actually set; the density of the air affected the shell's flight, and, what was more, affected different projectiles by different amounts; when all these, and other, effects were isolated and analysed it became neces-

sary to develop mechanical calculators which could be fed with the 'raw' information of range and bearing from the director and with information about wind, humidity, temperature, tilt and wear of the individual guns, and correlate all this to produce sets of data for each gun in which every possible error had been accounted for.

All that was left was to fire the guns, and even this brought problems. Simply shouting 'Fire!' at the gunners would have been a source of error in itself, since the varying reaction of the gunners would have resulted in a ragged broadside at best, so that centrally controlled firing by electricity became the normal method. The director layer would fire the guns by a push-button or foot pedal as his sight bore on the target – or almost, for the roll of the ship had to be taken into account. If the layer fired when his cross-wires were exactly on the target the guns would miss, since, due to human reaction time and the firing interval – the time between pressing the button and having the shells leave the muzzles – the roll of the ship would have moved the guns through an appreciable arc. The gunlayer had to judge the roll and fire as the cross-wires approached the target; later, even this element of art was taken from him and controlled mechanically by a gyroscope.

Lastly came the question of damaging the target, and here every navy was quite certain that it, and it alone, had the best answer. The period between 1885 and 1905 saw the most intense research into the allied problems of armouring ships and defeating the armour. The chilled-point Palliser shell reigned supreme as long as wrought-iron armour was the standard by which its efficiency was measured, but improvements in armour left it behind. Early attempts at mild-steel armour were moder-

routine tests of Palliser shell when, by accident, a wrought-iron plate was left in front of the armour target. Shells which had previously failed to damage the target now completely penetrated it, and English proposed placing a wrought-iron cap on the point of the shell; one was made and fired and passed through the plate. But authority was, for some reason, unimpressed, and no further action was taken at that time.

It remained for the Russian Admiral Makarov to revive the idea of the cap in 1890, but it was another five years before capped shells gained general acceptance, after which they became the standard method of attacking armour. The fitting of a blunt soft metal cap over the point of a shell may seem a peculiar idea, but experiments showed that its effect was to spread the force of impact on to the body of the shell instead of concentrating it on the point; it also supported the shell during the first stages of penetration and, at the velocities

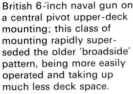

British 6-inch naval gun on a central pivot upper-deck mounting; this class of mounting rapidly super-seded the older 'broadside' pattern, being more easily operated and taking up much less deck space.

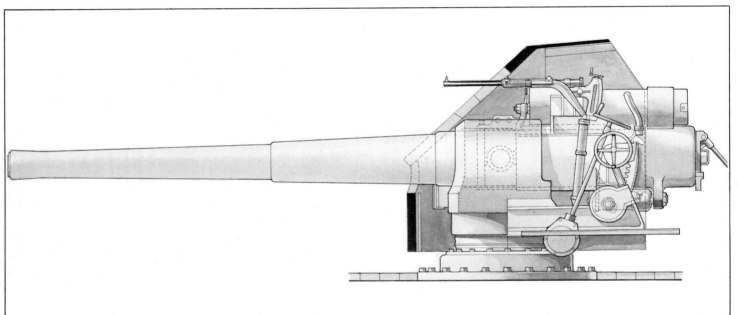

ately successful, but since this type of protection was prone to shatter when struck, the first major step was to compound armour – a wrought-iron back-plate faced with a hard steel plate. This was followed in 1889 with nickel-steel plate, and in 1890 by 'Harveyized' or carburized steel in which the front face had been made immensely hard by raising the carbon content. To pierce these new types of protection, steel shells were developed, the first really successful model being the French Holtzer shell of high-carbon nickel chrome steel.

However, it became necessary to increase striking velocities in order to penetrate, even with these projectiles, and it was found that there was an upper limit of velocity at which the impact of the shell was so enormous that the point and head shattered without penetrating. The solution to this had inadvertently been discovered some years previously by a Captain English of the Royal Engineers. In 1878 he had been supervising some

(Right) The operation of a naval rangefinder.

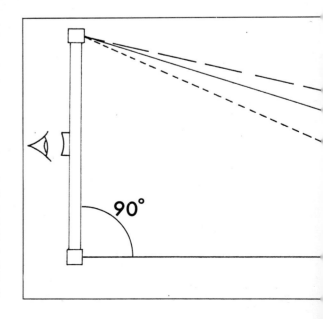

90°

and pressures involved, acted as a form of lubricant to ease entry of the shell point into the plate.

Originally the anti-armour projectile was a steel shot—a solid slug of metal—but in 1895 the Wheeler-Stirling Company in the United States introduced a 'Semi-armour-piercing Shell' with a small charge of high explosive initiated by a base fuze. This was capable of passing through a Harveyized plate two-thirds of its calibre thick and detonating behind the plate. This was soon followed by the Firth-Stirling 'Rendable Shell' which carried 2·5 per cent of its weight as high explosive and could penetrate its own calibre of armour. These immediately became the projectiles demanded by the world's navies and by 1905 the plain shot was obsolete.

Unfortunately there was one slight cloud on the horizon. The British Admiralty specification demanded a shell which would pass through its own calibre of hardened plate at a striking velocity of 2,000 feet per second and at 'normal'—i.e. striking exactly at right angles to the surface of the plate. This 'normal' angle requirement was a hangover from earlier days when attack was assumed to be a relatively close-range affair with flat trajectories. But by 1905 naval guns had extended their range to the point where the shell would be descending and would inevitably strike the target at an angle. The shell-makers (who were all private companies—there were no Government factories capable of producing piercing shell) tried to interest the Admiralty in oblique attack, but the official view was that if a shell could do well at normal, it would doubtless do as well as could be expected at oblique. In vain did the makers point out that the design of a shell for oblique attack needed to be somewhat different; the proof specifications remained unchanged and the makers had perforce to develop what one later described as 'specialized instruments for the attack of plate at normal'.

Unless action was likely, between-deck guns were dismounted and brought inboard, both to improve stability and for their own protection against the action of the sea. This illustrates a typical method of slinging the gun out of its mounting and anchoring it to prepared chocks on the gun deck.

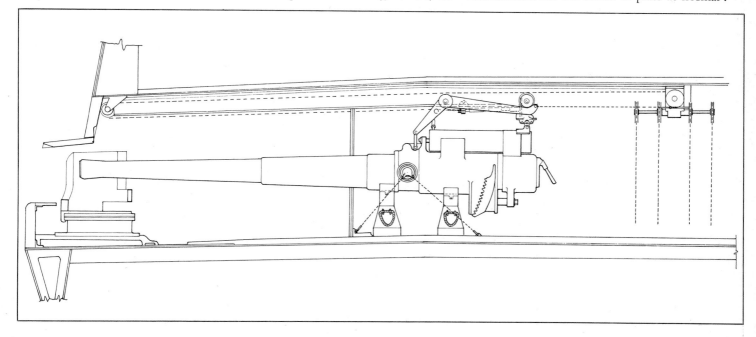

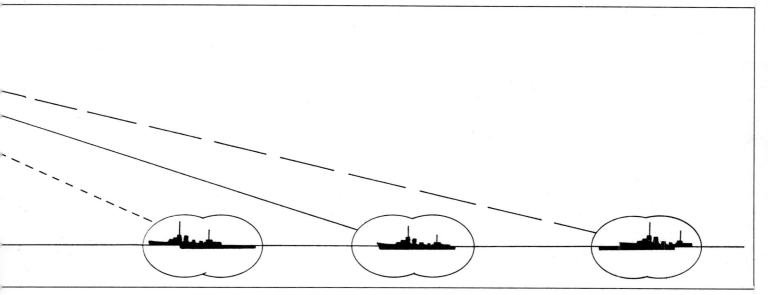

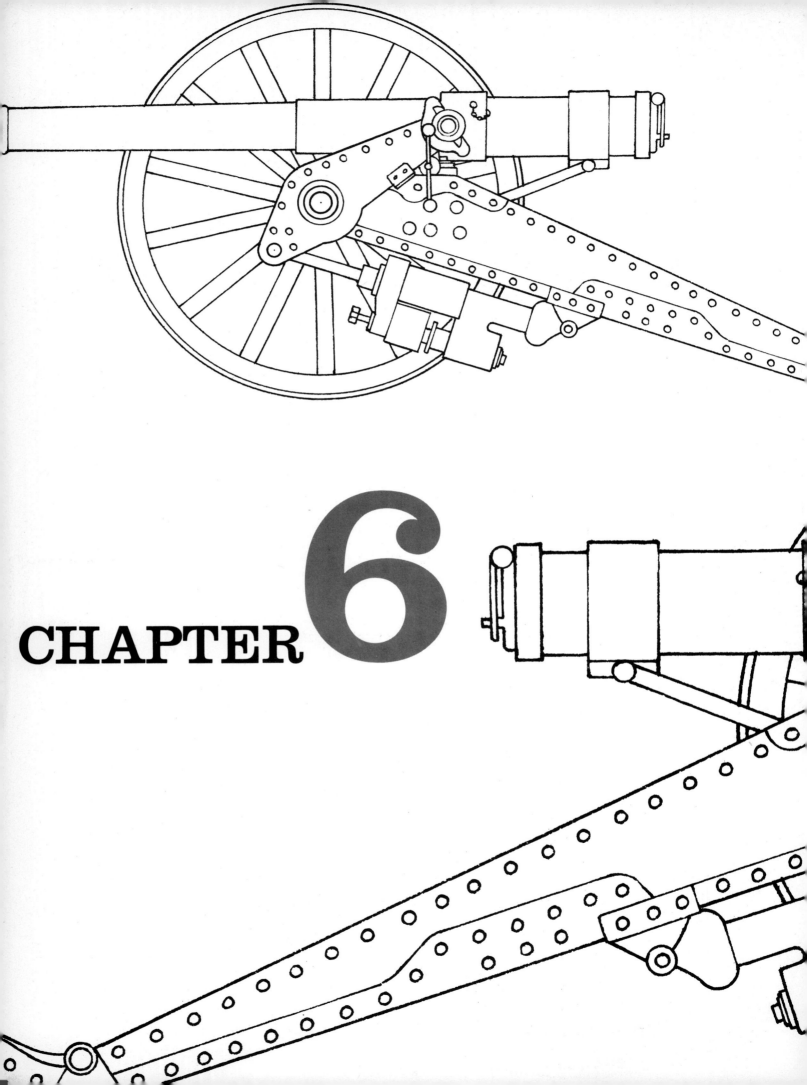

CHAPTER 6

Battle is Joined

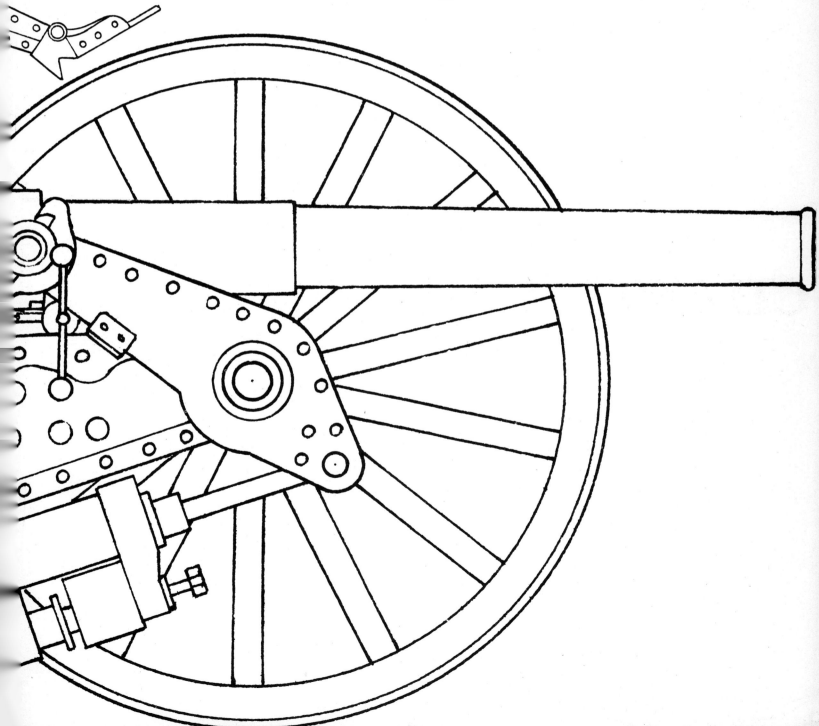

When the world went off to war in 1914 the artillery of the various nations was much of a muchness; field and horse artillery used weapons about three inches in calibre firing shrapnel shell to about 6,000 yards, the principal difference between the two branches lying in the weight of the weapon. There might also be field howitzers with the forward troops, about 105-mm to 120-mm calibre, again shrapnel firing but probably with a small proportion of the new high-explosive shells. Siege howitzers of larger calibre existed in small numbers, and between the extremes were a scattering of long-range guns in the 5-inch/155-mm class. Without exception armies were equipped to fight a war of movement; the Franco-Prussian, Boer and Russo-Japanese Wars all pointed to this. And, as the first weeks of the First World War were just that, it seemed that the forecasters had been right and each army was satisfied that they had equipped themselves with the right combination of weapons for the task ahead.

Then things began to go awry. The mobile war slowed down and, in effect, became a siege, demanding more and more and heavier and heavier ordnance. The other phenomenon which aston-

in the first 15 months of the war 2,750 guns were either destroyed, worn out or captured.

Similarly the French field-gun ammunition stock was 1,700 rounds per gun, 1,300 of which were ready for use and the balance held in arsenals as component parts to be assembled on mobilization. The plans called for wartime production to run at 3,600 rounds per day – a little more than three rounds for each gun at the front. But some batteries were blazing off hundreds of rounds a day, and after only a month of war the stocks were so depleted that there were only 500 rounds per gun on hand. Immense efforts had to be made to turn out shells at production rates undreamed of in pre-war days; by January 1915 600,000 per month were being produced, but still the demand increased; by mid-1917 the monthly production figure reached the incredible height of 7 million rounds.

The French were by no means alone in this; by the spring of 1915 the British Army was rationed to four rounds per gun per day, with the tacit understanding that even these were not to be fired unless absolutely necessary. The Austro-Hungarian Army was caught flat-footed in the middle of a decision to change its field howitzer and had sufficient of neither the old nor the new. And the Germans, though probably better prepared than anyone else, were nevertheless taken unawares by the sudden demands for more and more of every-

During the First World War the British Army demanded large numbers of heavy howitzers. The first models were made by rebuilding other weapons, but this 8-inch Mark 6 was designed and built by Vickers in 1915 and became the standard model for the rest of the war.

ished the armies was the rapid wastage of guns and the incredible consumption of ammunition. The French Army began the war with 5,556 75-mm guns and were of the opinion that such a vast number would see them through the war; consequently there were no plans for production during the course of the war. In fact their losses were so startling that within two months the gunmakers were being called on to produce more, and

thing that could fire or be fired.

The demand for replacement guns and vast quantities of ammunition was soon reflected in the quality of material appearing at the front. The French raided every possible source for weapons, and large numbers of antiques without recoil systems, ex-fortress guns, and even cast-bronze mortars last used in the Siege of Sevastopol were impressed for service. The German Army was

A British 9·2-inch Mark 10 railway gun opens fire.

bar, and filled with anything which was available and likely to explode. One defect of making shells of poor-quality material was that in order to give a sufficient margin of safety they had to be made thicker, leaving less room inside for the explosive. This deterioration in performance led to hurried research into the possibility of filling something else into shells other than the standard TNT (which was anyway in short supply); the Kaiser Wilhelm Institute of Berlin was asked to examine various chemical substances to see if they could come up with an alternative filling, but after their Professor Sackur was killed while experimenting with a cacodyl oxide mixture in December 1914, enthusiasm for strange substances waned.

However, the Germans had made a considerable propaganda victory as well as a tactical one with their attack on the Liège forts on the Belgian border. These works, laid out by Brialmont, the foremost military engineer of the turn of the century, were considered to be impregnable to any sort of attack which a field army could bring against them, and there was a good deal of astonishment when they were reduced and taken in a scant three days. It appeared that some enormous guns had been brought against them by the Germans; but such rumours were obviously nonsense. One contemporary journal, *Arms and Explosives*, was quite definite about it all, and included the following piece.

less concerned with weapon supply than with ammunition; instead of shells being slowly forged and machined and perfectly finished of high-grade steel, they were now rapidly cast from pig-iron, pressed from stock commercial steel

First introduced in the Boer War and retained through the First World War this 4·7-inch was the result of the marriage of a naval gun with an elderly carriage dating from the days of the rifled muzzle-loaders.

The 4·7-inch, limbered up and ready to travel.

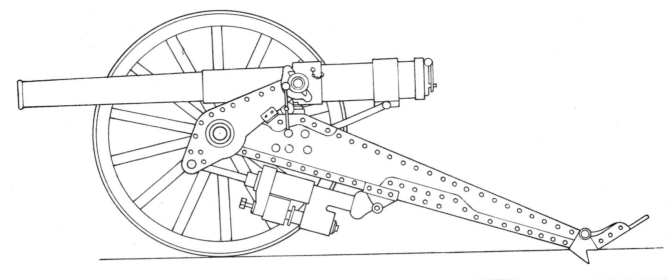

A British 12-inch gun Mark 9 on railway mounting, photographed at the Elswick Ordnance Works, Newcastle, before leaving for France in 1916.

'In *The Times* of the 21st ult., the military correspondent refers to the reported existence of a 28-centimetre German howitzer throwing a projectile of 345 kilogrammes to a distance of 10,600 metres. The calibre translates into 11 inches and the weight of shot into 760 lb. Remembering that a 12-inch gun weighs about 70 tons, and the mounting about as much again, a total of 150 tons is arrived at for firing an 850-lb projectile. An 11-inch howitzer firing a 760-lb shot sounds impossible . . . the notion of moving such a mass otherwise than on rails requires an active imagination. Stability of the mounting . . . would also present difficulties . . . which appear insuperable. The mere idea of loading a 760-lb shot puts a further strain on the imagination. Tall specifications . . . are a natural consequence, but unless they will bear strict examination they must be dismissed as an unnecessary attack on the confidence of our troops.'

The fact of the matter was that the weapons in question were not 28 cm or 11 inches in calibre; they were 42 cm or 16 inches; they were not firing a 760-lb shell but an 1,800-pounder; and far from weighing the estimated 150 tons, they weighed a mere 43 tons in firing order. Two howitzers existed, manned by Kurz Marine Kanone Batterie 3, and the impact of their enormous shells echoed round the world.

The Japanese, during the Siege of Port Arthur in 1904, had produced two batteries of 28-cm howitzers, largely to be able to shell the anchored Russian fleet, and movement of such monsters on

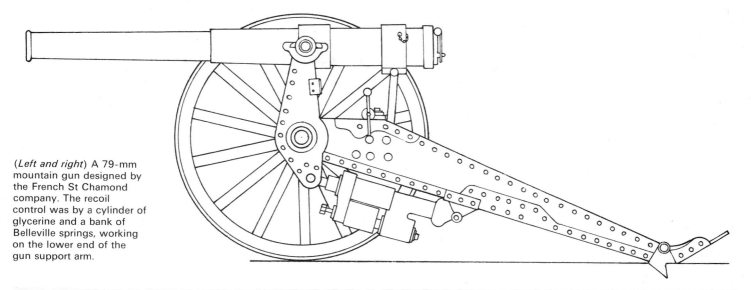

(*Left and right*) A 79-mm
mountain gun designed by
the French St Chamond
company. The recoil
control was by a cylinder of
glycerine and a bank of
Belleville springs, working
on the lower end of the
gun support arm.

HM King George V watches
the loading of the 14-inch
railway gun on 8 August
1918. At the left of the
King is General Sir Henry
Horne, Commanding 1st
Army, and on the right
Major General E W
Alexander, GOC Royal
Artillery, 1st Army. The
round was·fired at 18 miles
range, obtaining a direct
hit on the railway lines at
Douai; the shot went into
Regimental history as 'The
King's Shot.'

to the field of battle had caused some astonishment. At that time Krupp of Essen had made a
30·5-cm howitzer for coast defence and had begun
to ponder the problem of making such a weapon
mobile; no matter which way the German Army
moved, it was going to be confronted with
concrete fortifications, and the provision of a
heavy weapon to deal with them would be a
distinct asset. A long-barrelled version of the
30·5 cm was tried, but its 800-lb shell was felt to
be insufficient to deal with concrete of the thickness anticipated, and a larger weapon of 42-cm
calibre was designed. This howitzer, known as
'Gamma', had a 16-calibre-length barrel and fired
a 2,100-lb shell to 16,000 yards. For movement it
was dismantled by a crane and loaded on to six
railway waggons; if a suitable line was available

it could be brought into position and set up in 36
hours, and weighed 175 tons when assembled.

On trial in 1911 its performance against fortifications was impressive, but its weight and
cumbersome system of transport and assembly
were against it, and Krupp's designer, Professor
Rausenberger, went back to his drawing-board to
start afresh. Shortening the barrel by two calibres
slightly reduced the range, but the lowered muzzle
energy permitted the mounting to be redesigned
in lighter form, and by 1913 the two 42-cm
Mörser L/14 were ready for issue. Although on a
two-wheeled carriage of more or less conventional pattern, they were dismantled for movement
on transport waggons and assembled by a simple
gun hoist in a matter of four hours or so. After
their début at Liège they were supplemented by a

An 18-pounder in action, manned by a New Zealand detachment.

A French 34-cm railway gun manned by US Coast Artillery personnel, in France, September 1918

number of Austrian 30·5-cm howitzers of similar type, made by Skoda, and these two patterns of howitzer proceeded to demolish every fortress which confronted the German advance in Belgium and France.

The lessons of these monsters did not take long to sink in. The British Army had been performing trials with a 9·2-inch howitzer of their own for some months prior to the outbreak of war, and promptly put this into production. This weapon had been designed by the Coventry Ordnance Works, a company formed early in the century by a consortium of engineering and steel companies, and they now began designing a 15-inch howitzer based largely on the 9·2-inch model but scaled up in size. Firing a 1,400-lb shell to 10,795 yards, this seemed to be a useful weapon, and the managing director of the company, a retired admiral, got in touch with former colleagues at the Admiralty in the hope that they would pass the word along to the right quarter and get the Army interested. Instead, the First Lord, Mr Winston Churchill, decided to adopt the weapon as a naval contribution to the war; he ordered another seven to be made, manned them with Royal Marine gunners, and sent them to France. Later, in 1916, the Navy felt that they had enough to do on the high seas, and presented the howitzers to the Army. By that time both the British 15-inch and the German 42-cm were showing their defect—a range of just over 10,000 yards was insufficient for the conditions of war which now existed on the Western Front. In 1917 the Germans withdrew their 42-cm from the field, and after trying to develop a lighter shell in the hope of producing a longer range, the British Army decided the exercise was not worth pursuing and simply retired the 15-inch to a quiet section of the front until it could be declared obsolete.

But the fashion for building big had taken hold; with mobile war put to one side, it seemed that the only solution was bigger and heavier shells to tear holes in the opposition, demolish barbed wire, wreck defences and generally steam-roller a way through the enemy lines. The first to see the quick solution were the French. The Schneider-Canet Company had been experimenting with putting heavy guns on railway mountings for several years, without raising much interest. Their original idea had been to develop a mobile coast defence weapon which could be moved about rapidly to cover a coastline with the minimum number of guns, and they had managed to sell a few to Peru and Denmark in the years before the war. Now they took the remaining stock models they had in the factory and presented them to the French Army as mobile heavy guns. The Army responded by gathering up all the heavy fortress guns they could find from distant points, and borrowing or begging elderly fleet reserve guns from the Navy, then presenting this heterogeneous lot to Schneider with a request that they mount them in similar fashion.

It takes time to design gun mountings, recoil systems, sights, carriages; all have to be carefully developed to deal with the stresses of firing and travelling. Neither the French Army nor Schneider-Canet were being allowed the time for a leisurely development programme. The principal difficulty lay in the production of recoil mechanisms; the guns which the Army had provided were all from

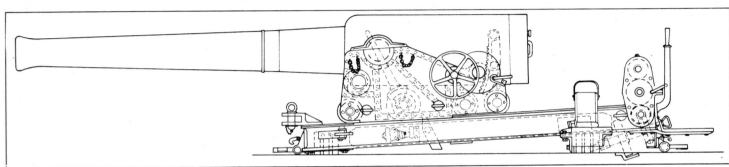

British naval 6-inch 80-pounder of 1882, the first of the 'new' breech-loaders which replaced the RML guns. It was mounted on the 'Broadside Sliding Carriage' and originated a long line of 6-inch guns in British service.

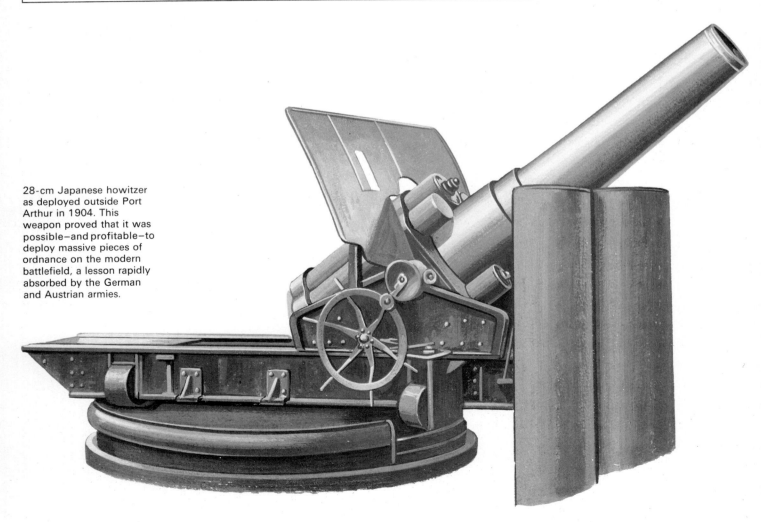

28-cm Japanese howitzer as deployed outside Port Arthur in 1904. This weapon proved that it was possible—and profitable—to deploy massive pieces of ordnance on the modern battlefield, a lesson rapidly absorbed by the German and Austrian armies.

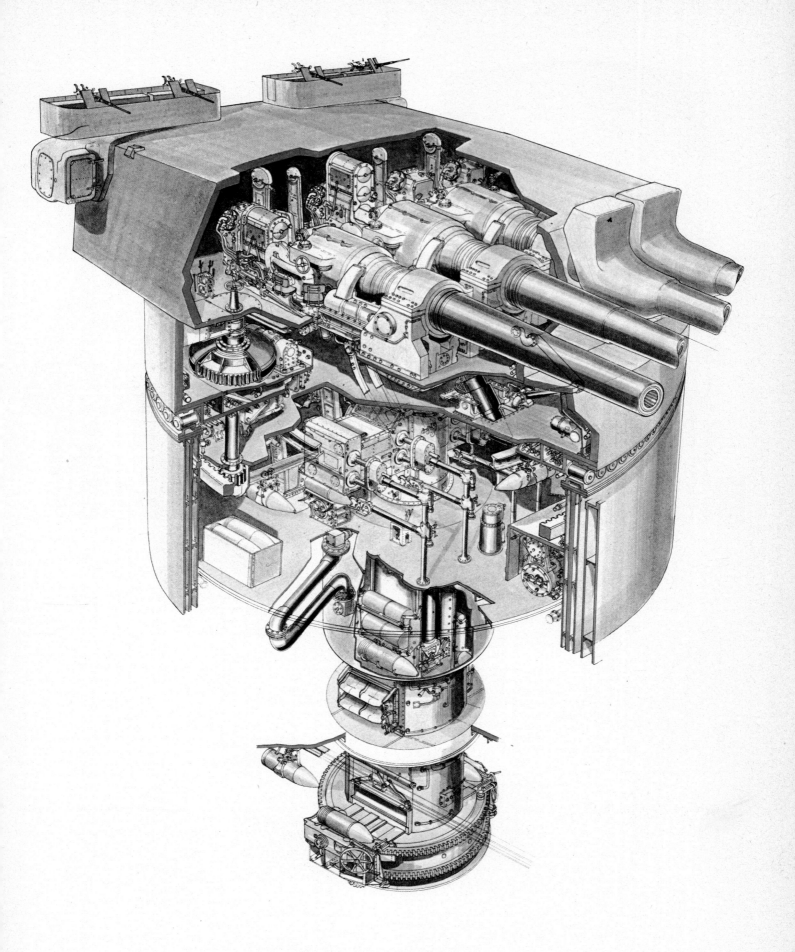

The four-gun turret of the *Richelieu*.

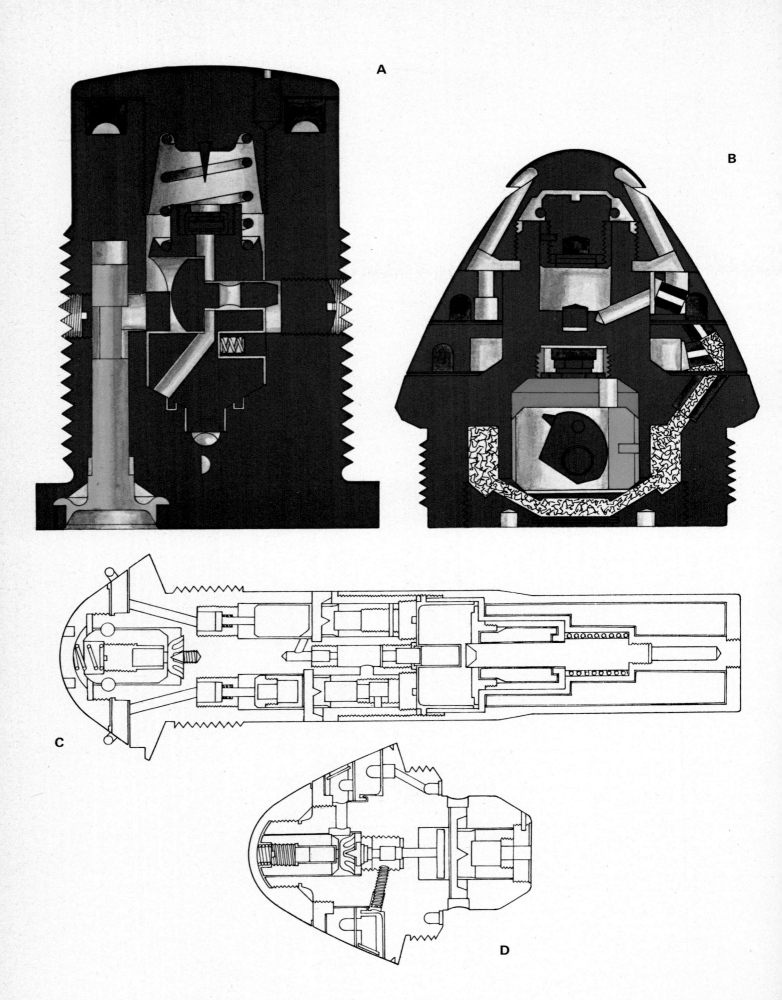

(*Opposite*) Fuzes, the key component of the shell.
(A) British No. 16 Base Fuze, 1916. Used with naval piercing shell, the cartridge explosion forced up the piston on the left; this allowed the central safety bolt to move sideways under the force of spin. When the shell struck, the central unit was driven forward to carry a detonator on to the needle, the flash passing through delay channels to fire the magazine in the head and thus initiate the shell filling.
(B) British Fuze No. 85, 1915, for anti-aircraft and field gun shrapnel shells. When fired the central detonator moved back, flashed, and lit the powder train which eventually fired the fuze magazine. Also, spin caused the firing pin to line up and if the shell struck before the fuze time had elapsed, then this pin caused the necessary action to take place.
(C) German Granatzunder 04, used during the First World War. A Krupp design which allowed the fuze to be set to give a delay after impact. Most complicated, and relies on several pellets of gunpowder which burn away during flight to unlock safety devices.
(D) German howitzer time fuze Model 1905. Similar to (B) above, it used a powder train for timing the shrapnel burst and also had a detonator and fixed firing pin in the base for percussion action.

(*Right, above*) A British 12-inch Mark 4 howitzer, showing the large earth box at the front; this was filled with 20 tons of earth to stabilize the weapon against recoil and jump.

(*Right, below*) Another view of the 12-inch howitzer, showing the loading cranes in place. This was one of the first field weapons to use a hydraulic rammer, pressure for which was provided by the recoil of the gun operating a pump which can be seen above the barrel and alongside the recoil system cylinders.

old-type carriage and slide mountings, with the somewhat rudimentary recoil system built into the slide. Once divorced from their static mountings the guns were devoid of any sort of recoil brake, and designing and manufacturing over a dozen different patterns to suit the variety of guns on hand would take years. So Schneider came up with a brilliantly simple answer which would never have stood a dog's chance of being accepted in any less fraught times: they dispensed entirely with the recoil system and mounted the gun straight into the railway mountings. These were simply massive side plates joined by cross-transoms and supported on a suitable number of wheels. The gun trunnions fitted into reinforced supports; when brought to the scene of battle, the mountings were lowered by jacks until the weight rested on a reinforced bed of girders between and alongside the track, all weight being thus removed from the wheels. In this position they were fired; the shock of recoil simply passed straight into the mounting and slid it back along the girder bed until all the force was absorbed. Since the recoil energy depends on the weight of the projectile *vis-à-vis* the weight of the mounting, the movement amounted to some three or four feet. After a few shots the gun would no longer be pointed at the target, and the gunners then had to jack it back on to its wheels and then, by means of hand cranks engaged with gears on one axle, move the whole mounting back to its original point, jack it down and open fire again.

As the French engineer remarked about the early motor-car gear-box, it was brutal but it worked. Inevitably the hammer blows of the recoil began to crack the side plates of the mounting after 300 or 400 shots, but by that time the barrels were worn to the point of inaccuracy anyway, and the whole weapon was simply put back on its wheels, dragged away and cut up for scrap. Although improved designs of railway mounting appeared in due course, these 'sliding mountings' remained in use throughout the war and, indeed, a number of them survived the subsequent peace to reappear briefly during the Second World War.

It was from this primitive beginning that the railway gun blossomed into the major innovation of the First World War. Using rather more elegant underpinnings and with the addition of recoil systems, guns up to 18 inches in calibre were produced and employed. Not all were entirely successful; the French, having been upset by the German 42-cm howitzers, demanded something bigger and better just to prove that they could do it. As a result a 52-cm (20·5-inch) howitzer was constructed; originally it was intended to put it into a non-recoil sliding mounting, but the recoil energy was too much for this sort of solution, and a proper design of carriage with recoil system had to be made, all of which took time. In the end two of these weapons were built, but beyond being used to impress visiting politicians they saw little employment. They fired a 3,640-lb shell, but the maximum range of only 14,600 metres was hardly worth the effort of emplacing them.

In the matter of long-range shooting the Germans invariably walked off with the prizes. One of their earliest endeavours was to emplace an ex-naval 38-cm gun known as 'der Lange Emil' on the Belgian coast and shell Dunkerque

One of the British 15-inch howitzers, manned by Royal Marines, with the 1,400-lb shell being hoisted to the breech.

An Austrian 30·5-cm howitzer in the Dolomites in 1917.

An example of an ammunition limber, this one for the British 13-pounder Royal Horse Artillery gun of 1903.

from about 40 km distance. At first the recipients were unable to credit this to a gun and were looking for airships, but examination of shell fragments eventually convinced them; in spite of air attacks and other retaliatory attempts, the Long Emil remained in business for most of the war.

But his efforts were overshadowed by the legendary 'Paris Gun' or 'Kaiser Wilhelm Geschutz' which shelled Paris from a range of 68 miles in 1918. This was the brain-child of one Doktor Eberhardt, an Austrian ballistics expert working for Krupp's. In 1916 he began contemplating the theoretical problem of launching a projectile at such a velocity that it would pass into the stratosphere; there, less affected by air resistance, it would be able to achieve a considerable range before it had to re-enter the atmosphere. According to his figures this would enable a range of 60 miles to be reached, and he took his calculations to Professor Rausenberger, his chief. Rausenberger agreed, and set to work on a design of gun to turn the calculations into reality. He also set about finding somebody who might be interested in shooting such a weapon. Since the German Navy was more used to operating heavy

'Max', the German 38-cm railway gun derived from a naval gun barrel, in position on the Western Front.

(*Opposite, above*) The French 40-cm Mle 15-16 railway gun joins in a barrage during the French attacks on Verdun in June 1916.

(*Opposite, below*) A British 8-inch howitzer, July 1916. This is one of the early, hurried, designs, using a cut-down and bored-out 6-inch gun barrel on a carriage put together from parts of a 6-inch siege carriage, a naval pivot mounting and traction-engine wheels.

guns than the Army, and since Rausenberger proposed using a naval gun as the starting-point of his design, he took the idea to Admiral Rogge of the Naval Ordnance Department. Rogge approved and it was then put before the Army; they too approved, and work began on building the gun.

The basis was a 38-cm naval barrel in a railway mounting, already in service as the 'Max' gun in small numbers. One of these was stripped and the gun barrel bored out to take a 21-cm calibre liner. This liner was much longer than the parent 38-cm barrel and had to be braced by a series of supports above the barrel to prevent a dangerous degree of droop. Before much else could be done, Hindenburg contacted Rausenberger and told him that he would be greatly obliged if the gun could be made to shoot 75 miles instead of the forecast 60; Hindenburg knew he was going to have to fall back in the near future, and a 60-mile range would not reach Paris—the only worthwhile target for such a weapon.

The gun was eventually finished and proof-fired in mid-1917 and work then began on preparing the firing site. This was selected in the Forest of

Gobain, 68 miles from Paris. Railway track was laid and reinforced concrete firing bases built, turntable rings set into the concrete, and camouflage prepared. But the gun programme had come to a halt due to the discovery that the shells designed for the gun were unsuitable. Fresh designs were drawn up and tried, and in the end it was March 1918 before the monster cannon was pulled into place over its prepared mounting bed, lowered down and bolted to the turntable and the wheels of the railway mounting removed.

The designing of the shells had not been easy. At the velocity and pressures involved the usual form of copper driving band would shear off before spinning the shell, and it was necessary to go back to Cavalli's design of the previous century and produce a shell with ribs on the body which would engage with specially deep rifling grooves in the barrel. Before opening fire the angle of elevation had been carefully calculated to reach the desired range at a specified muzzle velocity, and in order to achieve this range it was vital that the muzzle velocity be constantly attained. But with the size of propellant charge used there was enormous erosion of the chamber and rifling with each shot, with the result that when the shell was rammed it took up a position some inches ahead of that taken by the previous shell. This meant an increase in the cubic capacity of the gun chamber, and it was therefore necessary to measure the length of ram for each shell, calculate the chamber volume, and then calculate the weight of propellant needed to reach the desired muzzle velocity of 5,000 feet per second.

Another minor problem was the basic one of knowing where the shells were landing—at 60 miles range a man with a pair of binoculars was of little use. According to reports which have never been officially confirmed—or denied—this was done by a network of German agents within Paris who reported via a variety of contacts across the country to the Swiss border, from whence the information was telegraphed to Berlin and finally to the gun battery, arriving some four hours after the gun had fired. This process, which must rank as the longest and most roundabout observer-to-gun link in gunnery history, was only used for the initial rounds; once it was known that the shells were falling in central Paris, that was good enough, and subsequent firing was continued on the same data.

In all, eight barrels were built, with the intention of firing them at 21-cm calibre until they were worn out—estimated to be after some 50 to 60 rounds—and then reboring them to 24 cm. Three mountings were constructed, and the first barrel was fired for three days before being withdrawn for relining. A second barrel was then installed but, again according to unconfirmed reports, this suffered a premature detonation of a shell after only a few rounds had been fired. It is believed that this was due to loading a shell out of its proper order; since the gun wore slightly after each shot, the shells were made with gradually

The German 77-mm Feldkanone of 1916, mainstay of the divisional artillery during the latter half of the war.

increasing calibre and serially numbered, and the story goes that a high-numbered shell was inadvertently loaded into a barrel still too small for it. In any event, the third barrel was installed and fired for a time. The first barrel, relined to 24 cm was then brought back into use and the records show a distinct falling-off in regularity and accuracy of shooting with this barrel.

To say that the world was astonished at this weapon was putting it mildly; and yet there was nothing about it which was unusual or unknown; any competent ballistician could have obtained the same results, provided he could have persuaded

his army to believe him. Sir Andrew Noble, designer of guns for Armstrong at the turn of the century, had produced a 6-inch gun firing at the then unheard-of velocity of 3,200 feet per second in 1885, and the Ordnance Committee had developed a smaller weapon producing 4,000 feet per second for ballistic studies in 1904, but the rapid wear of these guns with the propellants then in use had ruled them out for practical use. It remained for the Germans to accept this drawback and turn high velocity to practical account.

By and large, the gun-makers during the war were simply called upon to produce more and

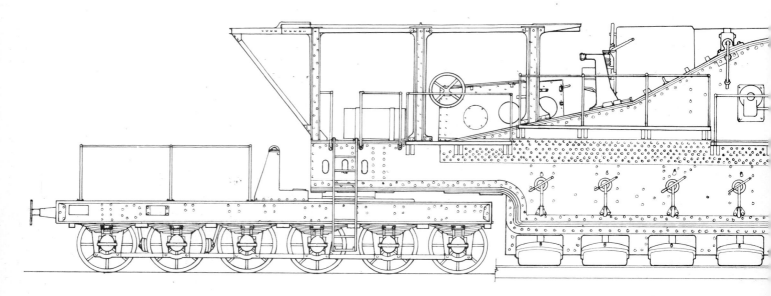

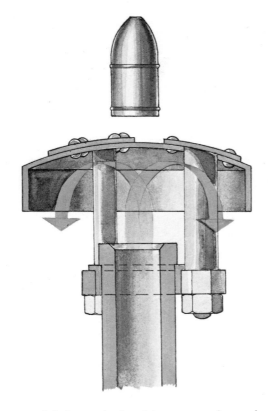

more of their standard articles or to make gradual improvements on proven designs; but in one or two fields there was little previous knowledge and the designer had a free hand to produce whatever he thought would do the job. One such field was that of anti-aircraft guns. As we have seen, this subject had been briefly discussed before the war, and one or two weapons developed. The British Army had been issued with a number of Maxim 1-pounder 'Pom-pom' guns, purchased during the South African War for field use, and now mounted on modified carriages so that they could fire upwards; the first of these had been tested in 1913 on the Isle of Wight, the target being a kite towed by a naval destroyer steaming at full speed off the shore. The French had a handful of their 75-mm

'Autocannon' on De Dion chassis, but the supply situation for the 75-mm was so precarious after the war began that it was a long time before they could provide any more. The Germans had the Krupp and Erhardt guns which had been developed in 1909, three of each on motor mountings and a dozen on horse-drawn carriages. The motor guns, with 1,000 rounds of shrapnel each, accompanied the field armies in 1914, while the horse-drawn guns, with 2,000 rounds apiece, were placed in strategic positions along the Rhine. To augment this number, captured French and Russian field guns were bored out to the standard German 77-mm calibre and provided with high-angle mountings enabling them to fire upwards at angles up to 60 degrees.

In Britain practically everything which could be persuaded to shoot upwards was tried at one time or another; eventually the 13-pounder horse artillery gun was provided with high-angle mountings, both lorry-borne and static, and became more or less the standard weapon. Since this seemed to work well, the 18-pounder was also tried, but this, due to the ballistics of the shell, was less successful, and a compromise design using the 18-pounder gun relined to 3-inch calibre so as to fire the 13-pounder shell with an 18-pounder cartridge was produced. Much more successful than such an unlikely combination might have been expected to be, it remained in use for many years after the war.

While the production of guns was difficult enough, the actual handling of them was even more involved. The target presented by an aircraft was like nothing which had previously been met; it moved fast and it moved in three dimensions. The nearest thing to this had been the problem of shooting at moving ships, and much of the coast gunner's expertise and technique was taken and modified. The principal problem was the fact that the shell took a finite time to get up to the aircraft, by which time the target had moved—and might have moved in any one of a dozen different directions at varying speeds. The

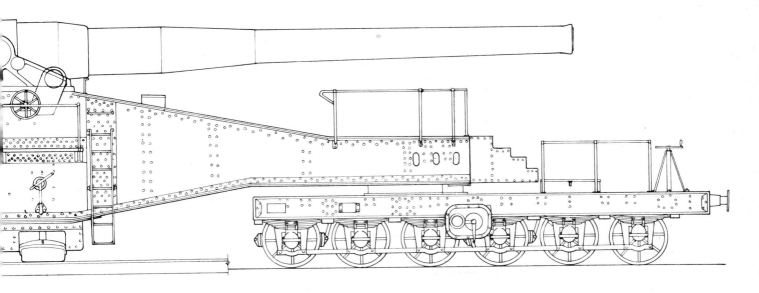

The first British anti-aircraft gun was the Maxim 1-pounder Pom-Pom, with its carriage modified to allow firing at high angles.

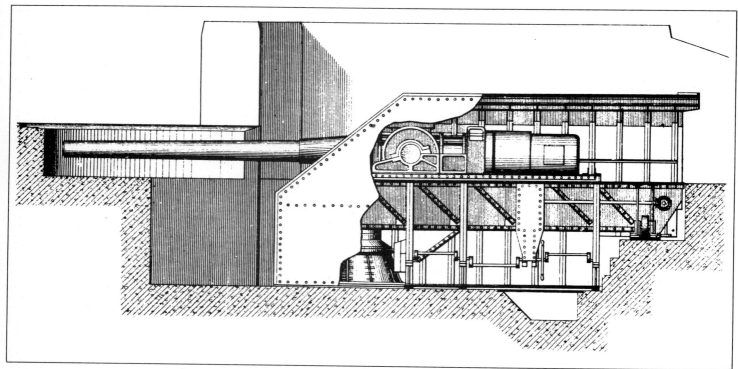

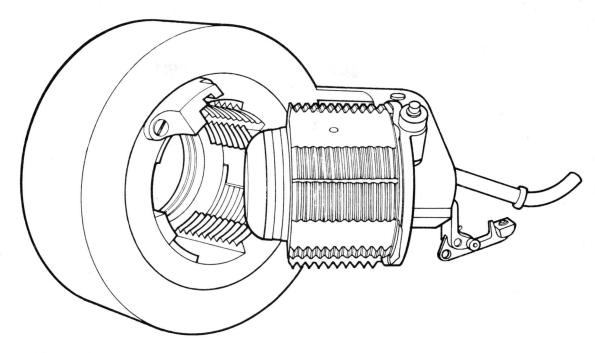

problem resolved itself into one of trying to guess –predict was the more elegant term–where the aircraft would be and then aim the shell at the estimated spot. The first attempts at this produced some highly ingenious mathematical concepts and some very complicated sighting systems and surrounded the gun with teams of calculators, but eventually the French came up with the solution which was accepted as being the only practical one and which has been used ever since. They concentrated the arithmetic in the centre of the gun battery, corrected the result so as to provide data for the widely scattered guns, and then told each gun what to set on its sights, what setting to put on the fuzes and when to fire. This cleared the calculators away from the guns, and also reduced the number of senior wranglers needed per battery, helped the rate of fire, and produced an acceptable answer. But in spite of increasingly sophisticated apparatus, by later standards anti-aircraft shooting was still in the wet-finger stage when the war ended.

Notwithstanding, the gunners made some remarkable scores. If the fire control and weapons were primitive, so, it must be remembered, were the aircraft. A post-war analysis showed that the German anti-aircraft guns had managed to shoot down 1,588 aircraft during the war; the British

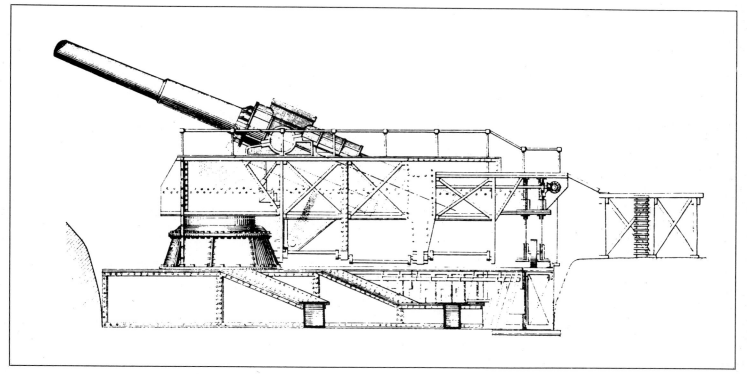

The original 'De Bange' breech mechanism. The screw passed through the 'ring carrier' and required half a turn to lock in place. The handle could be folded outwards to give extra leverage. Inset is the sectioned view showing the 'mushroom head' in front of the breech screw, and the resilient pad between the two.

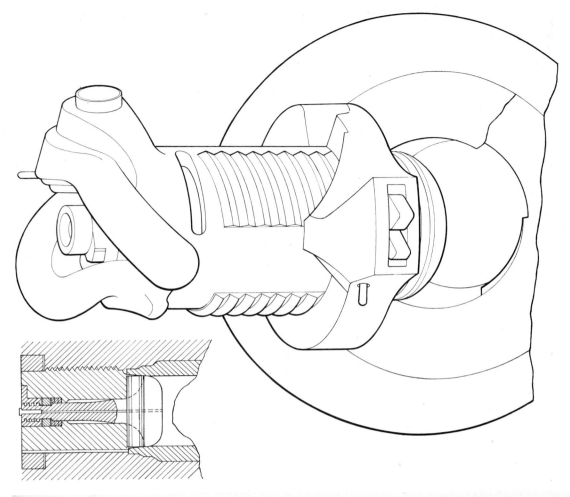

An 18-pounder in the mud during the Battle of Ypres. (*Imperial War Museum*)

The French answer to any problem was usually the 75-mm, Model 1897; here it is on the De Dion Auto-canon mounting as an anti-aircraft gun.

An 18-pounder after a direct hit; the Somme, 1918. (*Imperial War Museum*)

found that in 1917 it had taken an average of 8,000 shells to bring down one aircraft, while by the latter part of 1918 this figure had been brought down to 1,500 shells per bird. The most astonishing set of figures were those of the American Army; although only in action for four months, with two skeleton gun battalions and two machine-gun battalions, the anti-aircraft troops managed to bring down 58 planes for an average of 608 shells each. Another interesting correlation is the number of guns in action compared with the number of aircraft brought down; the Germans in 1918 brought down 748 aircraft with 2,758 guns, while the French in the same period claimed 220 planes with 800 guns; this works out at about 3·5 guns per aircraft in each case.

Another interesting design sphere was the alliance of the internal combustion engine with the gun; the early anti-aircraft gun designs had shown that such an alliance was possible, but it must be remembered that the anti-aircraft philosophy in the early days of the war was to turn the weapon out like a fire engine and chase the poor unfortunate aviator. The provision of self-propelled guns was thus solely to give the necessary pursuit capability, and not to endow guns with the capability of tactical movement with an army. This latter concept did not make an appearance until the tank arrived on the battlefield, for the simple reason that horse-drawn guns could move as fast as the foot soldier. Admittedly the guns and the infantry could both move faster than the original tanks, but methods of crossing

143

bullet-swept ground attracted one or two experimenters who felt that this might be the answer to moving guns across shell-holed country to keep up with an attack going forward; several battles on the Western Front had failed due to the fact that the advance had outrun the range of the supporting artillery, and guns could not be pushed forward fast enough to give the necessary constant support.

As behoved the country who first put the tank in the field, the first idea came from Britain, but it was not a particularly good one. Using a chassis based on components of the current tank, it was in effect a track-mounted platform on to which a 60-pounder 5-inch gun could be run; the gun wheels were then removed and the whole affair driven to wherever the gun was required. Once arrived, the wheels were replaced and the gun run off to be brought into action in the usual way. It appears that one or two experiments were made in firing the gun from the carrier, but this was not accepted as a workable proposition. In any event the idea was too complicated for what it achieved; an ordinary Holt tractor could pull a 60-pounder

An early example of mobile anti-aircraft artillery is this British 3-inch gun on a 4-ton Peerless lorry. Notice the stabilizing outriggers with screw jacks and wooden blocks, taking the firing shock off the vehicle springs.

A British 12-pounder coast gun mounted on a 'travelling carriage' for the movable armament of forts.

The British 3-inch once again, this time on a towed carriage which was the forerunner of similar designs in every country in the world.

During and after the First World War anything which could shoot was a potential anti-aircraft gun. Here is an 18-pounder, with fire control apparatus, being tried out in the role.

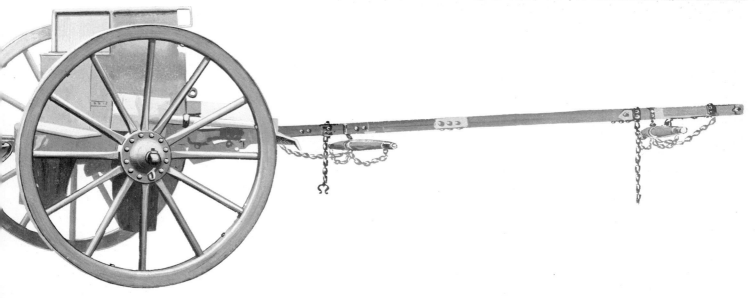

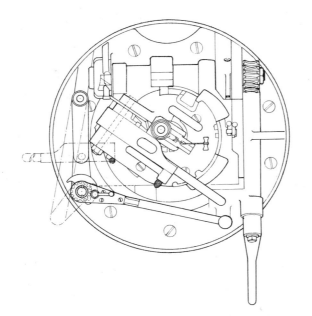

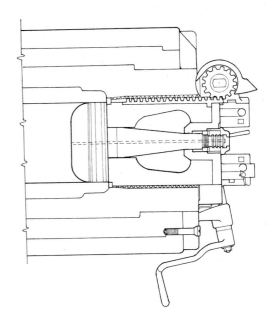

(*Right*) A British 9·2-inch breech mechanism of the early 1890s, entirely hand operated. A ratchet bar on the left was used to un-screw the block, after which it was withdrawn and swung open by turning the hand crank.

(*Far right*) A sectioned view of the 9·2-inch breech mechanism. The block was hollowed to save weight and carried a 'translating rack' on its right side which engaged with a revolving gear to withdraw the block.

An experiment in self-propulsion was this US 4·7-inch Model 1920E on a Christie carriage. The enormous blast deflector and muzzle brake are of interest. The gun was not very promising and was abandoned for an improved design.

faster than the tracked carrier could move and seemed to be equally efficient at cross-country work. In the end the gun carriers were converted into armoured supply carriers and spent their remaining days hauling rations and rum to the front-line infantry.

The French approach might have been foreseen: they put the 75-mm Model 1897 into a tank, placing it in an armoured box on top of a tracked chassis. Their idea was more in keeping with the self-propelled-gun concept as it is known today: to bring the guns up quickly to support the attack. But the design was bedevilled by poor execution, the carriers being mechanically unreliable, and when used in support of Nivelle's offensive on the Chemin-des-Dames they failed to live up to their advance billing. After this the official policy was to provide a vast number of light and mobile two-man tanks armed with machine-guns, and heavy-gun tanks were abandoned. But the gun-makers were still interested in the idea, and as a private venture both Schneider and St Chamond produced self-propelled guns which were adopted

in small numbers towards the end of the war. In these the idea of armoured protection was abandoned and they were simply tracked chassis capable of taking the guns anywhere; moreover the barrels mounted were those of the heavy weapons–155 mm and 240 mm–which were normally difficult to move about the battlefield. The success of these designs was instrumental in keeping the self-propelled-gun idea alive through the post-war years.

At sea the principal interest lay in the question of what would happen when the two great fleets of Germany and Britain came together, but it was not until the much-disputed Battle of Jutland that the question came near settlement. In this affair some fundamental truths were discovered: the necessity for efficient sealing between magazines and turrets was underlined when in the first phase of the battle the British battle-cruisers *Indefatigable* and *Queen Mary* both received hits on their turrets, which flashed down the hoists into the magazines, detonated the contents and sank the ships. Of the 2,000 men manning these two ships, only 18 survived the blasts. The German battle-cruiser *Derfflinger* almost suffered the same fate when, also struck on the turret, the explosion flames passed through the hoists and killed almost every man in the ammunition-handling rooms but failed to ignite the contents of the magazine.

From the ordnance point of view the greatest lesson from Jutland appeared to be the failure of the British armour-piercing shells to achieve the results that were expected. This was largely due to the defect which the shell-makers had been arguing for years–that at the long ranges involved the shells were striking at angles up to 20 degrees from normal and failing to penetrate. Those which did penetrate appeared to do less damage than their German counterparts, due partly to the fact the German shells carried a rather higher proportion of their weight in explosive and partly because the angled impact placed an unusual stress on the base end of the British shells which prevented the base fuzes from operating properly. The whole story will probably never be known, since it has become obscured over the years by accusations and counter-accusations, but it seems probable that the principal fault lay in the retention of old designs of capped shells using lightweight caps, efficient at normal but less efficient at oblique attack. After the battle a Naval Shell Committee was formed to try and improve

pressure hull. The 12-pounder gun was favoured for this sort of attack, as the shell contained sufficient explosive to tear a sizeable gash in a submarine's thin plating and the gun could keep up a good rate of fire to improve the chance of hitting. Another approach was to provide a heavy and high-capacity projectile which, in the event of a near-miss, still provided a big enough explosion to damage the target; numbers of 7·5-inch and 11-inch howitzers were specially designed to

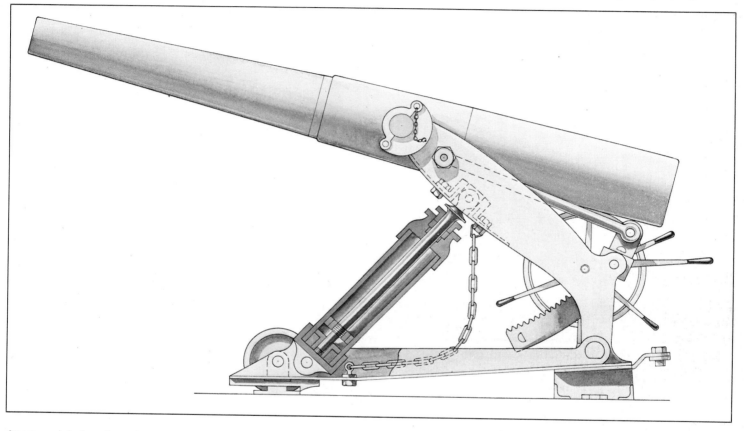

An unusual design of naval gun mounting used with the German 87-mm gun of the late 1880s. The recoil lifts the top carriage, which then pulls the operating piston out of the recoil cylinder.

matters, and a piercing shell with a heavier cap, proofed at oblique attack, was issued in 1917, but since there was never another major fleet action its efficiency was never put to the acid test.

The greatest amount of design activity in naval circles was devoted to dealing with the submarine. From being a dubious joke before the war it had blossomed into a prime threat, and it became imperative to try and develop some method of combating it. While the depth charge dealt with the deeply submerged target, various methods of artillery attack were studied in order to attack the surfaced or barely submerged target. Firing a normal type of shell was a chancy affair, since the curved nose usually caused the shell to ricochet from the water and pass over the submarine, and flat-nosed shells were developed to counter this. Next came special 'diving' shells, some with flat or concave noses, some with flat discs secured to the front, so that on striking the water they took up a predictable underwater trajectory and could be aimed short of the submarine to travel through the water and then detonate on striking the

throw shells with impact and hydrostatic fuzes to short ranges, so that a direct hit would function the impact fuze or, in the event of a miss, the hydrostatic fuze would detonate the shell as it sank through the water and thus still have some effect on the submarine. A more unusual technique was the use of stick bombs, resembling the missiles being used with trench mortars at the time, consisting of a heavy spherical bomb on a long rod; the rod was loaded into the muzzle of a conventional gun, a special blank cartridge loaded into the breech, and the result was a sort of depth charge with an impact fuze which would function either on hitting the submarine or the water alongside it. The variety of missiles and techniques evolved well reflects the concern which the U-boat had aroused.

CHAPTER 7

Intermission

For all the lessons in gun design which came out of the war, by far the greatest area of learning was in the handling of the guns themselves. It will be recalled that the armies of 1914 were designed for open warfare, and the events of early 1915 when the war came to a halt and settled into its siege form caused considerable changes in tactics. One change was that the guns and gunners vanished from sight; the South African and Russo-Japanese Wars had shown that the days of standing wheel-to-wheel in full view of the enemy were over, but although this had led to the development of indirect shooting, there was still, in 1914, a body of opinion which held it demeaning to hide from an enemy—not only must the guns support the infantry, they must manifestly be seen to be doing so. This attitude is shown in the famous action of the Royal Artillery at Le Cateau during the retreat from Mons in 1914. General Smith-Dorrien took the decision to stand and fight at Le Cateau in order to check the German advance and give his battered corps time to regain its breath, hoping to hold them off sufficiently to allow him to slip away in the night before the German reserves could be brought up and a full-scale attack launched at him. By the time his orders reached the artillery they implied that this stand at Le Cateau was to be a fight to the finish. Accordingly the artillery commander, General Headlam, ordered the guns to come out of their covered positions and take up their stand in the open, where they could be seen by the infantry and where the fact of their support could never be in doubt. The subsequent action raged from six in the morning to three in the afternoon, and in

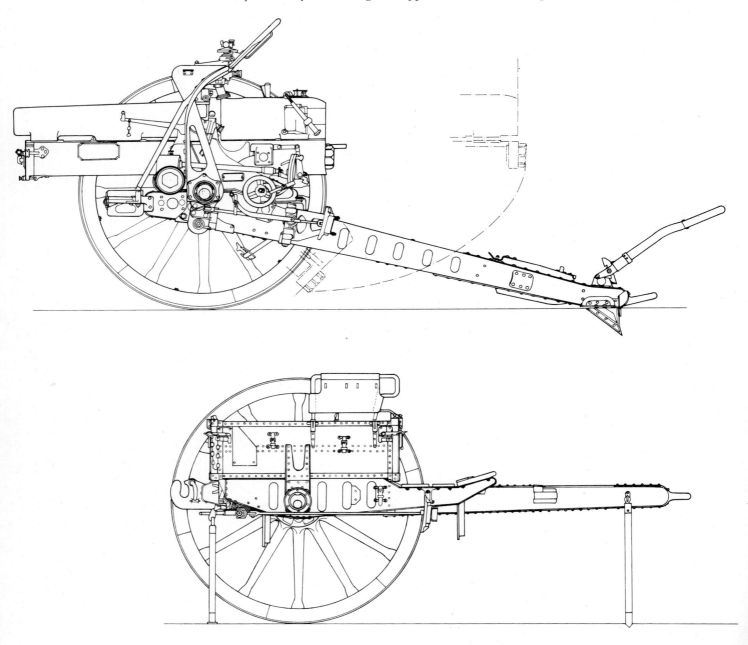

spite of severe losses among the guns and gunners the advancing Germans were fought to a standstill, so battered and bemused that with few exceptions the surviving British troops were able to withdraw during daylight without German interference.

Le Cateau was the last time that guns were to be exposed in such a fashion: the price was too great. From now on the digging of gun-pits and erection of camouflage became subjects of intensive study and practice. With the stagnation of trench warfare came the massing of guns to try and hammer a way through the opposition. Unfortunately the general shortage of ammunition precluded any great strides at first. With the French reserve down to less than 500 rounds per gun and the British reduced to three or four rounds per gun per day, little in the way of evolution could be expected, but some small moves were made. For the first time, guns were allotted bearings and ranges on

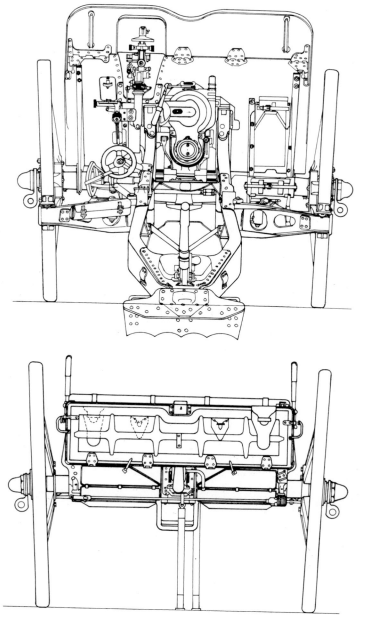

(*Opposite and below*) Section drawings of the British 4·5-inch howitzer of 1909, and ammunition limber.

which they were left laid at night, ready to open fire instantly to block an enemy movement against the front line. Attempts were made to harass enemy transport routes and forming-up areas during the night when activity in these areas might be expected. There were also thoughts directed to how best to locate and attack the enemy's artillery, and tentative experiments in the use of aircraft to try and gain some knowledge of what was happening across the front line and where there might be some likely targets. During the winter of 1914–15 barbed wire made its appearance, and it was soon realized that no infantry attack could hope to succeed against wire unless aided by artillery, which in turn led to a wide variety of trials and experiments to discover the best combination of projectile and fuze for removing this new obstacle.

The winter lull had the advantage that it allowed some stock of ammunition to be built up, and in March 1915 the British Army made its first serious attempt to break through the German lines at Neuve Chapelle. The site had been selected with care; the frontage of the attack was no more than 2,000 yards wide, held by six German companies, and three infantry brigades were detailed for the attack. To support them every available gun was collected together to give a strength of 246 13-pounder and 18-pounder field guns, 36 4·5-inch howitzers, 38 4·7-inch guns, 24 6-inch howitzers, four 9·2-inch howitzers, one 15-inch howitzer and five 6-inch guns. Against this, the German artillery consisted of but 24 field and 36 heavier guns. This meant that the British had amassed one gun for every four yards of front, an impressive figure; but the impression is diminished a little when it is realized that the ammunition available was, at the most, 500 rounds per gun. The heavier weapons had even less, the 15-inch, for example, having but 35 shells for the whole battle, so that it was vital that the artillery fire be kept to the minimum and that some sort of decision reached before the guns ran out of ammunition.

The German position was a single system of trenches of simple character protected by two or three rows of barbed wire, and in order to cut these entanglements for the infantry, 90 18-pounder field guns were carefully sited where they could open fire with shrapnel at relatively short range, this having been found to be an effective method of dealing with wire. The remaining guns were to engage in a general bombardment for 35 minutes before the infantry attack began. The field guns, except those allotted to wire-cutting, were to bombard the German trenches, while the heavier weapons spread their fire over the rear defences, the village of Neuve Chapelle itself, and the near-by German artillery batteries who might be expected to come to the aid of their infantry when the attack began.

Once the 35-minute bombardment ceased and the infantry began their advance, the artillery fire was planned to break off from its bombardment of the front-line trenches and concentrate on the rear defence and the village; and after a short interval,

as the advance closed up to the village, the fire would again change to a box of fire around Neuve Chapelle in order to prevent reinforcements coming up. It was the first example of artillery planning and it succeeded remarkably well; unfortunately two 6-inch howitzer batteries were late in arriving and thus they were unable to shoot any ranging rounds previous to the battle but had to take the range and bearing for their targets from the map. Map shooting was a little-known art at that time and the fire of these two batteries was, on the whole, somewhat ineffective, but on the rest of the front the whirlwind bombardment disrupted communication and so thoroughly un-

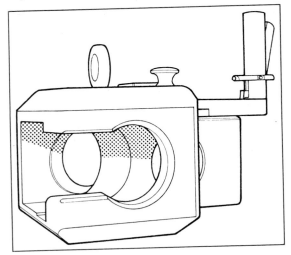

(Far left) The Mark 1
18-pounder of 1914 and
(below) the Mark 4 of
1920, illustrating the
improvements in design due
to the lessons learned
during the First World War.

nerved the defenders that the attack was able to go through with little opposition. But after the initial breakthrough there was a lack of reinforcement and the battle was only a limited success.

The fact that the attack had succeeded in its initial stages was taken as evidence that the prime function of artillery was to wreck the defences; the failure of the attack at the northern end of the line, where the two 6-inch batteries were firing, was read as being due to their having failed to do sufficient damage to the German trenches. What was overlooked was the more important effect on morale of a sudden storm of 35 minutes of intensive shellfire on to the unprepared heads of the

The result of this was seen at the Battle of Festubert in the following May; the preparation for the attack consisted of a 48-hour bombardment, and the failure of the attack was attributed to insufficient destruction wrought by the guns. Moreover, the increase in the width of attack for later battles was not accompanied by an increase in the amount of guns available, so that the guns were now thinly dispersed: instead of the one gun to four yards of Neuve Chapelle, the Battle of Loos in September 1915 had one gun for every 23 yards of front, and in order to make up for this deficiency the duration of the initial bombardment had to be increased. In contrast to this drawn-out

(Left) A modern application
of Krupp's sliding block,
from the American 75-mm
pack howitzer shown on
pages 190 and 196.

(Right) The 37-mm French
'Trench Gun' which started
a fashion for lightweight
infantry guns during the
First World War. It was
virtually a scaled-down '75'
and fired a useful high
explosive shell.

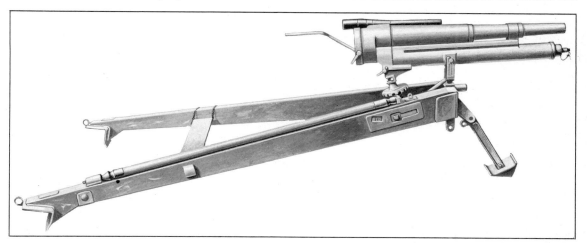

One of several patents
taken out by the Holt
Caterpillar Manufacturing
Company in 1917–18 to
cover their proposals for
self-propelled guns; this is
basically a field carriage
provided with tracks and a
driving motor.

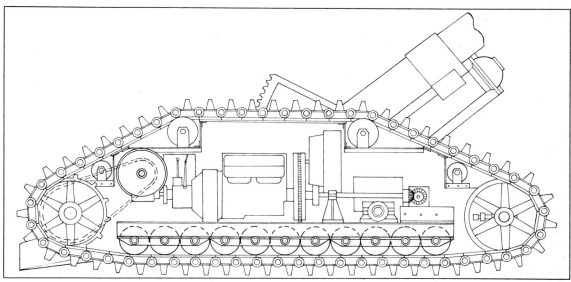

poor unfortunates in the front line; by the time the bombardment stopped and the infantry attack began, the defenders were so demoralized and confused that they were incapable of putting up any sort of coherent defence. As a result the Allies were led into a policy which, as Field-Marshal Lord Alanbrooke later said, 'had as its ideal the complete destruction of all defences irrespective of the fact that such a procedure entailed the sacrifice of all surprise and the creation of new obstacles to our advance in the shape of shell-torn areas'.

bombarding, which merely served to alert the enemy that an attack was coming and allow him to concentrate his reserves, when the German Army attacked Verdun in February 1916 their preliminary bombardment, from approximately 1,400 guns, howitzers and mortars, was intensely concentrated, being relatively short and extremely violent.

In spite of this sort of lesson, the Allies clung to their policy of long bombardment, culminating in the enormous affairs of 1917. At Verdun, for example, the French had 2,300 guns and fired

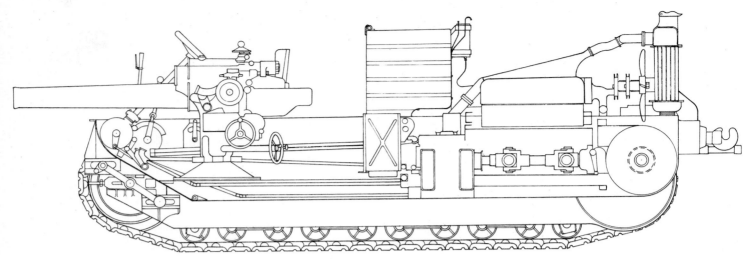

(Top) Another early patent for self-propulsion. This actually pre-dates the arrival of the tank, having been granted on 21 February 1916.

(Above) The US Army purchased the design of a 24-cm howitzer from France in 1918. Numbers were modified for use as coast defence weapons in Hawaii, but they were not particularly successful.

120,000 tons of ammunition – 360 railway train loads – in a fortnight-long bombardment. It was the introduction of the tank, with its promise of a new method of beating down wire and neutralizing machine-guns without having to shell them for days beforehand, which helped to break the stalemate and bring some sense of proportion back to the planners.

It will be recalled that in the early days of the war the German Army had called upon the Kaiser Wilhelm Institute for the Advancement of Science to see if they could devise some filling for shells to replace high explosive and also to provide cheap cast-iron shells with some worthwhile effect. Stemming from this, experiments began on filling shells with offensive gas in the hopes of incapacitating the defenders while allowing the attackers, suitably protected, to overcome them. The first gas-shell attack was by the German Army against the Russians at Bolimov in January 1915, in which a number of 15-cm howitzer shells containing xylyl bromide were fired. Owing to the extremely cold weather the liquid failed to vaporize properly and was ineffective, but in spite of this setback the idea of using poison gas had caught the German imagination and it was eventually introduced on the Western Front in the form of gas-cylinder attacks. In order to deliver it at longer ranges and in order to be less dependent upon the wind (which generally prevailed in a direction unfavourable to the Germans) the search for suitable artillery shells to contain gas continued, howitzer shells being eventually issued in late 1915. It was the French who made the most intensive study of the subject, as befitted their status as recipients, and they appreciated that the defect of the German shells was that they were basically high-explosive shells into which a small container of gas had been inserted. It seemed obvious to the French that, since artillery shells were small and carried a small payload, it would be better to increase the proportion of gas so as to get the most benefit from the number of shells fired. To have any effect at all it was vital to swamp the enemy with gas, and shells with small payloads could not do this unless impossible numbers of guns were called in. They therefore developed shells in which the major proportion of the filling was gas, with only enough explosive to crack open the shell and allow the gas to disperse. With these shells, filled with phosgene, they gave the Germans a severe surprise at Verdun, and very quickly the gas shell built on these lines became the standard article on both sides of the line.

The only remaining area of research was to produce a gas which would give the greatest effect for the amount of payload carried, and many and varied were the compositions tried. One tried by the French was prussic acid, a poison deadly enough in the laboratory but lamentably ineffective in practice where the volatile gas quickly dispersed in the air. The Germans, with their highly developed chemical industry, were generally the innovators where new gases were concerned; their fillings were known by code names as 'Blue', 'Green' and 'Yellow Cross', names arising from the markings on the shells. Blue Cross was an arsenical smoke, Green Cross a phosgene mixture and Yellow Cross the most deadly of all, mustard gas. Blue and Green were often fired together in barrages, since Blue Cross would penetrate some types of gas-mask and cause the wearer to sneeze; thinking he could be no worse off without a mask

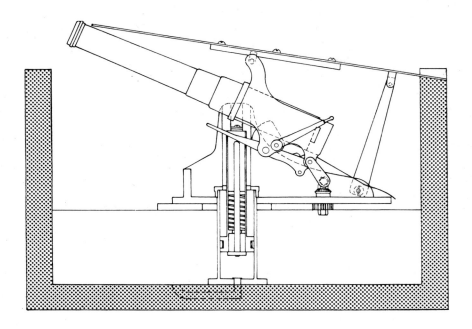

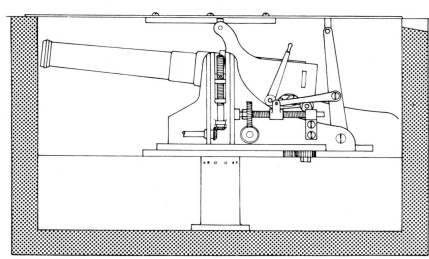

he would remove it and take a deep breath of the lethal Green Cross which he had not realized was there. Mustard of course not only attacked the respiratory system but was also lethal when it landed on the skin; but it was a persistent gas, and fired into an area it would contaminate the soil and water and form an effective barrier for the advance of troops; so much so that a bombardment with mustard gas was often received with mixed feelings by the Allied troops. On the one hand it meant some danger from the gas, but on the other it generally assured them that they were unlikely to be attacked for a few days.

As the war drew to a close the gun designers were at work preparing a completely fresh generation of weapons. It must be borne in mind that the guns with which the war opened were, for the most part, the first generation of modern ordnance. As the war had progressed so the demands for more range and greater power had given rise to improvements in pattern until, in the end, although the guns bore the same names, they bore little other resemblance to the guns of 1914. For example, the British 18-pounder Mark I of 1914 weighed 2,821 lb, elevated to 16 degrees and fired to a range of 6,525 yards. The 18-pounder Mark IV of 1918 weighed 3,116 lb, elevated to 30 degrees and ranged to 9,300 yards. The last inch of performance had been wrung out of every design by successive improvements, and there was little hope of being able to obtain much more without completely fresh designs of guns; but until the designers could get their ideas off the drawing-boards and into production, the gunners had to make do with what they could get; and nowhere was this more apparent than in the United States.

When the USA entered the war in April 1917 their stock of artillery was low by European standards: 600 3-inch field guns, 60 4·7-inch medium guns, and a handful of 3·8-inch, 4·7-inch and 6-inch howitzers of varying antiquity. With the

An example of the type of invention which proliferated during the First World War is this disappearing mounting complete with roof shield. Like most of the inventions, it was stronger on ingenuity than on mechanical likelihood.

An interesting experiment from the early 1930s; a weapon with interchangeable barrels, one a 75-mm high-velocity gun and the other a 12-cm howitzer, both using the same breech mechanism. This is a British model; similar designs were tried in France, but none was ever adopted for service.

The British 3·7-inch pack howitzer, last of a long line of guns developed for mountain warfare, and capable of being rapidly dismantled and carried on mule-back.

An American experimental 3-inch anti-aircraft gun, the Model 1923E, photographed during its trials. It was one of a series of weapons developed during the 1920s.

declared intention of putting 2 million men under arms by the end of 1918, this collection of weapons had to be fleshed out from somewhere; orders were given to manufacturers and plans drawn up for the building of more factories, but merely stating an intention is not providing weapons. As the Chief of Artillery, General William J Snow said later: 'The American people must realize that you cannot order a gun on Friday and have it delivered on Monday—you may get it in a year or a year-and-a-half.' In order to provide the first elements of the American Expeditionary Force in France with guns, a number of 75-mm Model

1897s were purchased from the French, but the prime intention was to provide the American Army with an American-designed and built gun. The 3-inch Model 1902, the standard field gun at the outbreak of war, was much like its contemporaries, a limited-elevation shrapnel-firing gun. In fact it was a very slightly modified version of the Erhardt 15-pounder which the British had bought in 1901. Recognizing by 1914 that it had fallen behind in the race, a new design was prepared of a gun on a split-trail carriage using a new hydro-spring recoil system. This eventually became known as the 3-inch M1916 and plans were put in hand for quantity production. But in order to simplify ammunition supply in the war zone, it was decided to standardize on the French 75-mm cartridge and shell, and the 3-inch M1916 was redesigned to become the 75-mm M1916, a process which, of course, delayed production while the redesign was done. In another endeavour to produce guns the British 18-pounder was adopted; this was in production in the USA for British contracts, and when these were completed the manufacturers were instructed to keep going, only this time for the American Army. Again, production was stopped while the gun chambers and barrels were redesigned for French 75-mm ammunition, and the gun then became known as the 75-mm Model 1917. Shortly afterwards, General Headlam, who was by then head of the British Military Mission in Washington, commented that they might at least call it the 'British 75', since they called the M1897 the 'French 75'; it was called this unofficially, but the name never stuck to the extent that the French one did.

The subsequent story of the setbacks and problems in American gun production make astonishing reading for anyone raised in the belief of American manufacturing expertise, but the trouble lay exactly where General Snow said. You cannot produce guns out of thin air, overnight; although the manufacturers did their best, only a handful of guns had come from the production lines before the war ended. Manufacture of the French 75 in

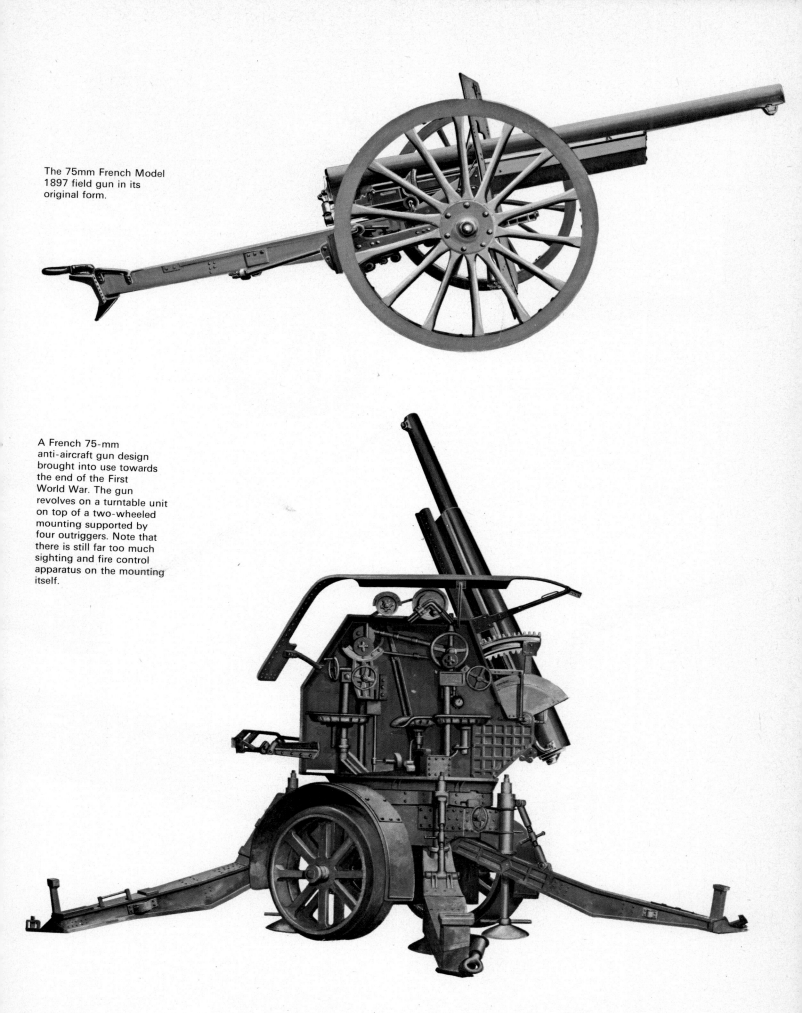

The 75mm French Model 1897 field gun in its original form.

A French 75-mm anti-aircraft gun design brought into use towards the end of the First World War. The gun revolves on a turntable unit on top of a two-wheeled mounting supported by four outriggers. Note that there is still far too much sighting and fire control apparatus on the mounting itself.

The British 3·7-inch mountain howitzer. Introduced in 1916, this was another jointed gun, with the muzzle and breech sections held together by a 'junction nut', so that it could be dismantled to produce two mule-loads for the gun and six for the carriage.

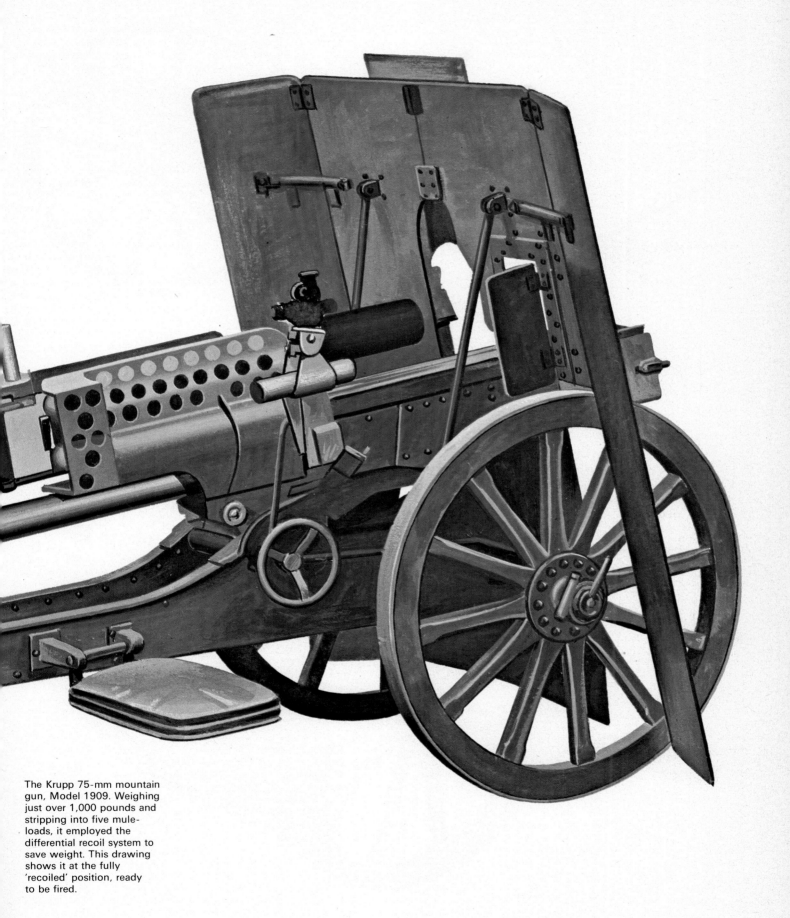

The Krupp 75-mm mountain gun, Model 1909. Weighing just over 1,000 pounds and stripping into five mule-loads, it employed the differential recoil system to save weight. This drawing shows it at the fully 'recoiled' position, ready to be fired.

The German 77-mm field gun, Model 1896. A combined Krupp/Erhardt design which formed the basis for numerous improved models for the following twenty years.

The British 13-pounder Royal Horse Artillery gun, designed after experience in the South African War and built as an amalgam of the best ideas put forward.

Even in 1919 the Americans were still building disappearing carriage guns. This 16-inch was the largest and last of the line; after its tests, shown here, it was installed in the Panama defences. It was scrapped in 1944.

the USA was bedevilled by the production of the hydro-pneumatic recoil system. It was impossible to get any information out of the French for a long time, and when they finally handed over the drawings they were petrified with fear in case some unauthorized person should see them and thus come into possession of their incomparable secret. When the Americans finally stripped one to pieces and examined it they found that the secret lay simply in the fact that it was a hand-fitted job with extremely fine tolerances; converting it into a mass-production proposition was incredibly difficult but was finally managed. By the Armistice, the American factories had produced 233 Model 1916 guns, 800 Model 1917 and one single French 75. The record in heavy guns was little better; persuaded by the French to accept their 155-mm guns and howitzers as their standard, none had been produced by the time of the Armistice, although the Dodge Brothers put up a ten-million-dollar factory for the production of recoil systems. For heavier weapons the US Army relied on supplies of 8-inch and 9·2-inch howitzers from the British, plus a number of railway guns constructed by mounting surplus coast gun barrels on to French Schneider sliding mounts.

When the war was over the armies of the world began to go through a thin time. While they were under no illusions about the effect of the 'war to end wars' their political masters were largely brainwashed and refused to countenance any expenditure on weapon development which was other than vitally necessary. Moreover, they had all finished the war with immense stockpiles of artillery; even the US Army came into the land of plenty when the wartime contracts finally began to come through – for it took almost as much time to stop production as it did to start it. Against this

background it is hardly surprising that little in the way of new equipment found its way into the hands of the soldiers during the 1920s and early 1930s, but within the limited budgets development work continued.

The anti-aircraft gun was one sphere of development; in the immediate post-war years there was a time when it looked as if the gun would disappear entirely. It became an article of faith that the natural enemy of the aeroplane was another aeroplane, and that guns had been simply a wartime stop-gap. Further development of aircraft would render them obsolete, it was held, and as a result there were very few anti-aircraft guns in existence by 1923. But the promised super-aircraft failed to materialize, the gunners took heart, and the back-room boys began working on designs against the day they might be needed. General Pershing had asked for a heavy anti-aircraft gun of 4·7-inch calibre to be able to down an enemy with one near miss, and in a slow and methodical fashion the Americans began work on this, together with a static 105-mm gun for harbour defence against air raiders. Work also began on a 105-mm howitzer for field use; in 1916 Colonel Charles P Summerall had been sent to France as an observer to report on artillery trends and tactics, and he had come to the conclusion that the French 75-mm was too small, and that a weapon with a heavier shell, more suited to dealing with the tank, would be of more use. He urged the design of a 105-mm howitzer, but he was overruled by Pershing's staff and the various French advisers. After the war a mission of inquiry was set up by the US Army to examine the performance of artillery and make recommendations for future armament. Their report called for 'a howitzer of about 105-mm calibre on a carriage permitting a vertical arc of fire from minus 5 degrees to plus 65 degrees and a horizontal arc

Britain's 18-pounder Mark 5, or the 'Birch Gun'; this version shows the gun barbette mounted.

A Vickers design of light-weight 70-mm infantry gun of the early 1930s.

of fire of 360 degrees . . . the projectile should weigh 30 to 35 pounds and should include both shrapnel and shell. A maximum range of 12,000 yards will be satisfactory. Semi-fixed ammunition and zone charges should be used . . .' To this broad specification work began, but it was to be a long time before it showed results; the fact that Colonel Summerall later became General Summerall, Chief of Staff, doubtless helped the project along.

In Britain the policy was laid down by the Master-General of the Ordnance in 1922 when he said:

'The reduction of the Army and the fact that we have only a small force available makes it all the more important for us to continue research and experiment in connection with new weapons, but it is not our policy to proceed to production of

new weapons for the equipment of the Army or for reserves . . . what we are endeavouring to do is to improve the various designs so that if we ever had to manufacture again we would manufacture guns up-to-date according to our ideas of modern design . . .'

At that time a new 30-pounder 3·9-inch gun on a split-trail carriage was being designed, a 3·6-inch anti-aircraft gun which had been developed in the latter months of the war was on trial, a 4·7-inch anti-aircraft gun was in the design stage, and a new 5-inch long-range gun was being laid out on the drawing-board.

In the following years a number of interesting designs were put forward, considered and then put aside for something else. A 3-inch gun to go on the carriage of the existing 3·7-inch mountain howitzer was proposed in 1923 after a request from

Another view of the Birch gun, with the gun at maximum elevation. It was some time before the hope of using the same gun for both ground and aerial targets was given up as impractical.

One of the designs which did not make it; a British high-velocity 6-inch gun of about 1930. Of interest are the loading tray, pivoting on the trail and shown here positioned at the breech, and the brakesman's seat and brakewheel.

India for a high-velocity gun for mountain troops; the 3·9-inch gun became a howitzer, was built and fired, and then abandoned; it was replaced by a design of 4·13-inch howitzer in 1925. In the same year came proposals for a twin 6-pounder gun for light coast defence against torpedo-boats and similar fast craft, and a proposal for a 3-pounder infantry gun. The most far-seeing design was for a self-propelled 18-pounder on a Vickers tank chassis; in response to an inquiry as to how it was proposed to use this sort of weapon, the Director of Artillery replied: 'It may be taken that the equipment may be employed for the support of a mechanized force, of cavalry, and of advanced, rear and flank guards; also for the purpose of rapid reinforcement of artillery already in action. The equipment is to be considered primarily as a ground weapon. Its anti-aircraft role is secondary, and concerns forward areas only.' The resulting

equipment became known as the 'Birch Gun', so named for General Sir Noel Birch, Master-General of the Ordnance and responsible for its introduction. Two versions were produced, one with the 18-pounder gun mounted in a tank type of turret, and one with the gun in an open barbette mounting so that it could elevate to 85 degrees and act as an anti-aircraft gun. Development had begun with the turretted model in 1925, the high-angle version being produced the following year, and in 1927 one battery in the Experimental Armoured Force was issued with the guns for extended trial. There is some evidence that a third design was proposed, and got as far as the wooden mock-up stage; this abolished the anti-aircraft function and made some minor improvements in the controls, but before it could be taken any further the experimental force was disbanded in 1928 and the self-propelled gun went into limbo. There was a cer-

The breech and sights of the US horse-drawn 105-mm howitzer M1, which also shows the balancing spring beneath the cradle which counterbalanced the weight of the barrel.

tain amount of empire-building going on at the time, and these new hybrids fell between two stools; were they guns or were they tanks? The Tank Corps were agitating for their dream of a self-contained private army with their own 'Royal Tank Artillery', while the Royal Artillery rather resented this implied take-over of their function. Add to this the financial starvation of the time, and the collapse of the self-propelled gun is hardly surprising.

In one country, however, there was less of a problem. Germany had been stripped of most of its wartime stock of armament and was left with little more than a token force. The great gun-makers were restricted by the terms of the Versailles Treaty, Krupp to making guns of above 17-cm calibre and Rheinmettal to making guns below that figure, and the number of guns they could make was strictly limited. With no vast stockpile of weapons to fall back on, this was a ripe climate for gun designers to get to work and start making plans for the future, and the 1920s saw a wide variety of weapons put on paper for future evaluation. Due to the Treaty restrictions, the gun-makers made temporary alliances with other companies to keep their design staffs at work and keep their expertise alive. Krupp sent their gun designers to work for Bofors of Sweden,

The Bofors 105-mm field gun ready to travel; guns with long barrels frequently have them pulled back in this manner and locked to the trail to prevent vibration placing excessive strain on the elevation gear.

A photograph from a pre-war German magazine, showing the 75-mm infantry gun Model 18 on manœuvres.

naval installations. Thus in England, the harbour of Portsmouth, which at the beginning of the century had boasted over two dozen forts and something in the region of 400 guns was reduced to less than half the number of forts and not much more than two or three dozen guns. Similar reductions were made all over the world, for technical advances in power control, fire control, and more powerful weapons capable of covering a greater area allowed the same weight of shells to be delivered from a smaller number of guns. In the 1890s, with fire control in its infancy and with guns of limited power on hand-operated mountings, it had been necessary to provide a vast number of weapons to be sure of getting enough shells into the target area to deal with an attacking fleet; technical improvement led to economy in material, which, as it happened, fitted in very well with the enforced economy in manpower.

In 1921 the Washington Conference had agreed that the upper limit for battleship guns was to be 16 inches; this effectively put an end to development of 18-inch and 20-inch guns which had been under way in Britain and the USA as the war ended. At the same time the Conference led to an agreement between Britain, the USA and Japan not to increase their fortifications in the Pacific area. In this case the Westerners were out-foxed; the most modern US defences in the Pacific area were those at Fort Drum, the famous 'concrete battleship' in Manila Harbour, where 14-inch guns were mounted in turrets. The remaining defences of the Philippines and Hawaii were largely 12-inch guns and mortars dating from the turn of the century. Britain had 6-inch and 9·2-inch guns in Hong Kong, also dating from the early 1900s, and

while Rheinmettal obtained control of a Swiss firm who acted as development engineers for the designs drawn up in Germany, and also obtained a controlling interest in an Austrian company who manufactured the finally developed weapons and sold them for export under their Austrian name.

The post-war years saw a considerable reduction in coast artillery all over the world. The wartime years had seen little use of coast guns, and, moreover, the restricted armies of the time found it impossible to man their vast coastal armaments. Much of the older equipment was therefore swept away and coast defence concentrated on essential

Another American experiment of the 1930s was this 75-mm divisional gun. It could function as an anti-aircraft gun, as shown above or as a field gun (*below*). It proved to be unsuccessful in either role, like so many dual-purpose weapons, and the idea was dropped.

A Bofors 80-mm anti-aircraft gun in the firing position (*right*) and 'folded up' (*below*) for travelling.

some equally elderly installations surrounding Singapore. The Japanese were well provided with modern turretted coastal installations all round Japan and Korea which were well in advance of any Western weapons in the area and, being of 16·1-inch calibre, they could comfortably deal with battleships, having guns of similar power as restricted by the Treaty. Also in 1921, the British Government decided to construct a naval base at Singapore in order to support future fleet activity in the Pacific, and such a base would obviously require defences. By some fluke, the area was not included in the Washington Treaty terms, so that there was no bar to construction. Due to political vaccillation, it was 1931 before work at Singapore got under way; none too soon, for by the following year Japan's expansionist policy was becoming all too obvious. However, the Army had been planning for the Singapore fortress since 1922, working out every detail down to the location of the last spare firing pin, developing new patterns of guns and mountings, fire-control equipment

A Vickers design of the 1930s, this 105-mm field gun demonstrates the articulation of the carriage necessary to give stability on uneven ground.

The British 3·6-inch anti-aircraft gun on tracked trailer; designed in 1918 it was much in advance of its times and formed a useful testbed throughout the 1920s, helping to perfect many features which appeared on later weapons.

and rangefinders, so that when the word was finally given to go ahead with construction, all the paper work was done and installation could begin immediately. In 1932 orders were given for plans to be drawn up so that work could commence by the end of 1934, but since the gunners had their plans all ready, work began immediately and by August 1934 three batteries of guns were already installed. The major armament was five 15-inch guns, the newest and most powerful coast defence guns ever used by the British Army, and these were backed by an impressive array of 9·2-inch and 6-inch weapons.

Then, in 1934, the Japanese repudiated the Washington Treaty and it was possible to modernize the guns of Hong Kong by fitting them on to new mountings giving greater elevation and range, as well as to install new batteries at Penang, Kilindini and Trincomalee to cover the various naval installations there. The USA, hamstrung with their isolationist policy, preferred to keep out

of the limelight and made no improvements in their defence in Hawaii or the Philippine Islands; their only major works since the end of the war had been the installation of a number of 16-inch guns, some on disappearing mountings and some on barbette carriages, to guard the approaches to the Panama Canal.

With the Greater East Asian Co-prosperity Sphere coming into being on the one hand, the other hand displayed the rise to power of the new German Third Reich. Taking one thing with another it became obvious to the more percipient soldiers that war was slowly becoming, if not inevitable, certainly more probable, and it behoved them to do something about it. So the rearmament programmes of the 1930s slowly moved into gear.

Since Germany was calling the tune, she was first off the mark, and in 1933 the first results of the design exercises of the 1920s began to come from the factories. New field howitzers and

infantry guns were the first products to be seen, but behind the scenes there was a great deal more being prepared. Krupp's designers in exile had returned to Essen with drawings of a new anti-aircraft gun which had been mulled over and redrawn until it was as perfect as it could be on paper; pilot models were built and tested, it proved to be as good in the flesh, and production of the famous 88-mm flak gun began. Even further in the background the planners had been working on designs of a long-range railway gun since the middle 1920s and this, too, was now turned over to the production engineers.

The performance of the Paris Gun, while impressive, had irked the German Army a little, since it had been controlled and manned by the Navy, and the Navy it seems, was never backward in reminding the soldiers about this in post-war years. (It is a minor mystery of the post-war years, incidentally, as to exactly what happened to the Paris Guns; they were never captured, no component parts were ever found, and interrogation of designers and workers after the war found them strangely reluctant to talk about the weapon to Allied inquirers. All the information about the Paris Gun which is known today has been painstakingly built up from information gathered here and there; there was never any official report or statement on the weapon from German sources.) The Army therefore decided to go about the design of a long-range gun which would put the Paris Gun in the shade and reassert the Army's rightful place, should the need ever arise in the future. Work began on a 21-cm gun with the barrel deeply grooved with eight rifling grooves to fire a shell fitted with splines or ribs which engaged in the grooves to deliver the spin. This system of con-

(*Top*) A pre-war picture of the first public showing of the German 105-mm anti-aircraft gun.

(*Above*) Wehrmacht parade, with 15-cm howitzers in the foreground. Notice that the barrels are uncoupled from the recoil system and pulled back on to the trail, thus relieving the elevating gears of stress while travelling.

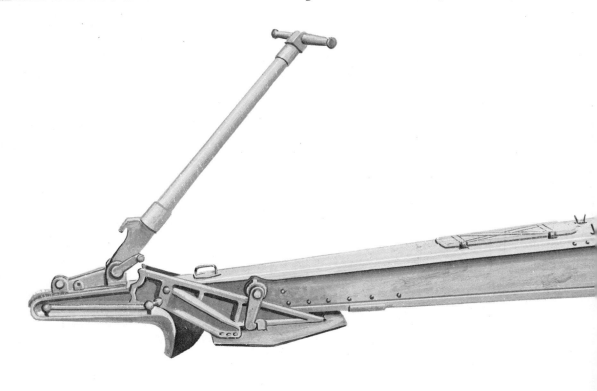

An example of long recoil aiding stability; a Bofors 90-mm field gun of about 1930.

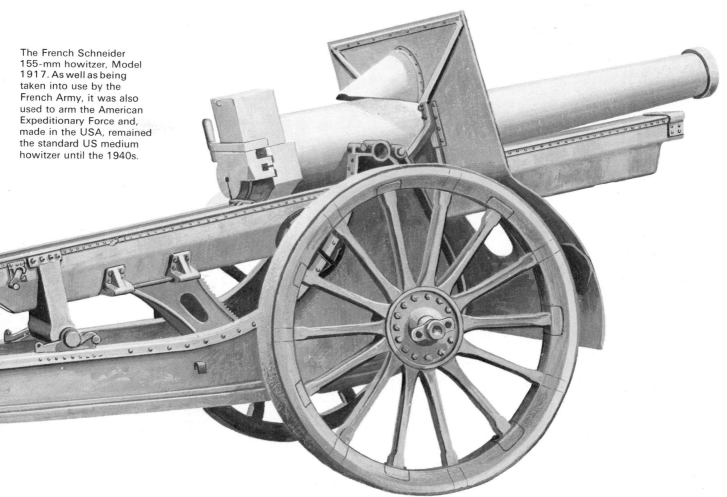

The French Schneider 155-mm howitzer, Model 1917. As well as being taken into use by the French Army, it was also used to arm the American Expeditionary Force and, made in the USA, remained the standard US medium howitzer until the 1940s.

Representative of British
Coast Artillery at its zenith,
a 6-inch gun at practice
near Plymouth.

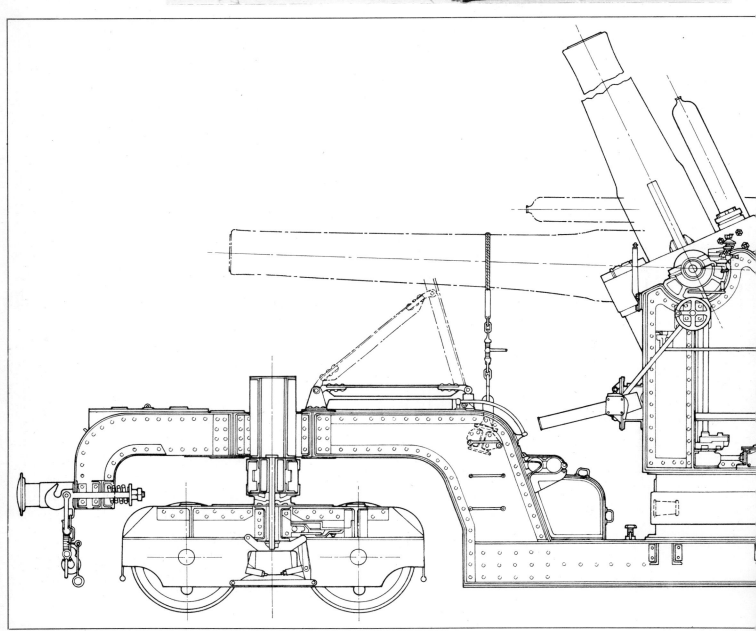

struction was resorted to since it was still unlikely that a normal copper driving band would be able to transmit the enormous rotational torque needed to spin a heavy shell at the velocity proposed.

In order to prove the point, a number of smaller barrels of 105-mm calibre were made with a variety of rifling patterns and fired at very high velocities. They confirmed the original idea and a full-scale 21-cm barrel, 109 feet long, was duly built and proved. A railway mounting was also designed, this being less of a problem than handling the ballistics, and it was more or less a simple steel box on wheels. But when it came to marrying the gun to the mounting some difficulties appeared. The enormous length of barrel had to be braced so that the muzzle end did not droop under the weight; this in turn meant a very heavy barrel structure which would be hard to elevate and depress. The use of some form of

hydraulic press to assist with the weight—an equilibrator, to give it the proper term—was common enough on smaller weapons, but on something this size—the barrel alone weighed 98 tons—it was hardly feasible. Therefore the gun trunnions had to be well forward, leading to a long overhang at the breech end, which meant that when the gun was elevated there was liable to be contact between the breech end and the railway track. To get round this the designers arranged for the whole of the mounting to be jacked up one metre from its wheels whenever it was to be fired, thus obtaining the necessary clearance between breech and ground. In addition, the mounting structure was connected to the wheel-bearing trucks through the medium of a hydro-pneumatic recoil mechanism, so that as well as the gun recoiling in the mounting, the whole mounting recoiled across its wheels; this 'dual recoil' allowed the recoil stroke of the gun

British 12-inch howitzer on railway mounting. Although provided with all-round traverse, it could only be fired within 20 degrees of the vehicle's axis, since it was not provided with stabilizing outriggers. Later versions corrected this fundamental defect.

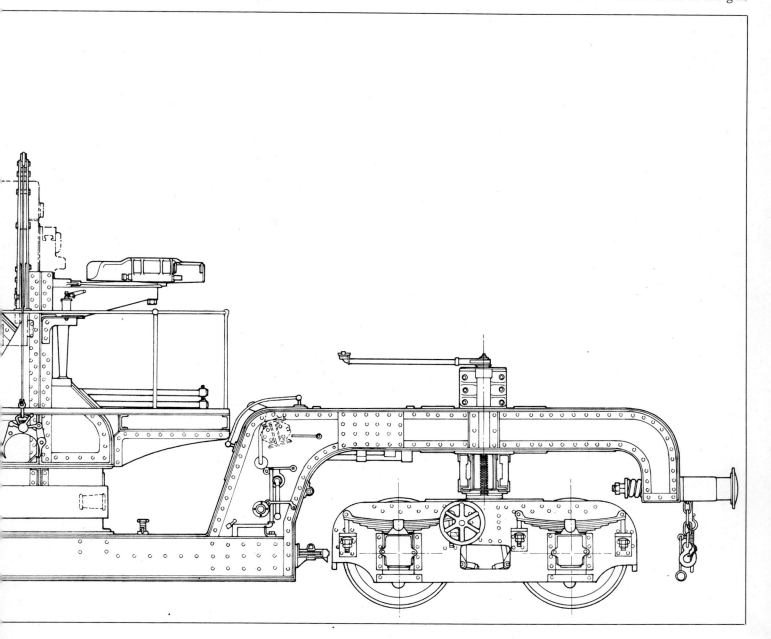

to be shortened and helped the breech-clearance problem.

The resulting weapon was known as the Kanone 12 (E): (12 for the forecast maximum range of 120 km, E for *Eisenbahnlafette* or railway mounting). It weighed 297 tons in going order and was issued complete with a special gun train which included air-conditioned ammunition waggons, crew coaches and a special firing track assembly. This allowed a 'T'-shaped track to be laid at the required site, so that the gun could be pushed on to the 'T'. The front truck units were then jacked up and rotated through 90 degrees and lowered on to the cross-stroke of the 'T', and the weapon could then be traversed by motoring the front bogies back and forth.

Other weapons were also being developed; more railway guns of more utilitarian character, heavy howitzers, anti-tank guns, heavier anti-aircraft guns, guns of every shape and size to fill the gaps in the resurgent Wehrmacht's inventory. The ammunition designers were called in, too, to try and evolve some new ideas which would allow more performance to be wrung from conventional guns; the full story of German artillery development during these years would fill several volumes.

Across the Channel, the British Government had at last seen the writing on the wall and had abandoned their long-cherished 'ten-year rule'; this, first predicated in the early 1920s, had assumed there would be no war for the next ten

(*Top*) Before a prototype gun is built, a full-size wooden mock-up is used to check that the detachment can actually serve it and to demonstrate what the finished article will look like. This was the mock-up for a proposed British 3·3-inch howitzer for Horse Artillery use in the early 1930s.

(*Above*) The experimental Vickers 105-mm howitzer of 1931. Although it was turned down, the carriage became the model for the immortal 25-pounder.

(*Right*) The British 25-pounder (3·45-inch) field gun-howitzer. With a range of 13,400 yards and a formidable anti-tank capability, it was undoubtedly the best all-round gun of the 1939–45 war.

A rare photograph of one of the prototypes of the 2-pounder anti-tank gun.

The production model of the British 2-pounder, showing considerable changes from the prototype design.

years, prohibiting weapon production. Sound enough when first announced, the credibility of the rule was eroded by constant restatement so as to extend the rule indefinitely and thus effectively prevent the military from ever acquiring new weapons. In March 1932 the rule was finally abandoned, in the face of Japanese advances in Manchuria and China, so as to allow work to begin at Singapore, and with the door thus thrust open work could begin on perfecting a number of weapons which had been mulled over during the lean years. A new field gun was needed to replace the ageing 18-pounder, and a new anti-aircraft gun was also of vital importance; numerous designs of both had been drawn, assessed, built in mock-up form, argued over and abandoned until there was agreement on what appeared to be the best solution. Even so the finances of the time had their effect: after discussing various possible field guns, such as a 20-pounder, a 22-pounder, a 3·7-inch and a 4·1-inch, the British Army ended up with a 3·45-inch 25-pounder largely because such a barrel could be fitted to existing 18-pounder carriages and thus economize on carriage production.

By the time all the designing and testing had been done in Britain, time was running shorter and shorter. In the normal course of events, when a gun design is produced a handful of weapons are built and exhaustively tested on an experimental range for a year or more; after this, a small number—enough, say, to equip a regiment—is manufactured to give the makers some experience in laying out their production line. It also allows a production model to be put into the hands of the troops enabling them to give it a more realistic test over a longer period to see if it would stand up to the rigours of active service in the hands of soldiers, a different thing to careful shooting on an experimental range in the hands of experts. All this can take three or more years before the final go-ahead to produce the weapon in quantity; indeed, it is a rule of thumb that from the first stroke of a pencil on a blank sheet of paper to the first general issue of a new gun can be taken as seven years. There was no time for this leisurely process in the mid-1930s, and as a result British designs were put into full production once it was seen that they worked, leading to what became known in the Army as the 'paper re-armament', since the guns went straight from their paper-plan stage to the troops with the minimum of in-between stages. It speaks volumes for the integrity and efficiency of the designers that the weapons survived this unprecedented step and were, in fact, among the best weapons ever produced.

While the gun makers were doing their part, the artillerymen were also working to perfect their systems of employment of these promised weapons. The First World War had seen changes in the handling of weapons, and the lessons learned at that time were exhaustively examined during the post-war years and used as a basis upon which to build. The techniques of indirect fire were sound, but in order to extract the best from them, and in order to provide that instant support which was the artillery's purpose in battle, fast and reliable communication was vital. Too often during the 1914–18 battles communications had broken down at a crucial time; telephone lines were cut by shellfire, runners failed to get through, pyrotechnic signals were primitive, and as a result the artillery had often been unable to use their power to its best effect. Every possible system had been explored, even to the use of carrier pigeons by observers; the best comment on this system came from the American Army, one of whose pigeons arrived at the gun position bearing the message: 'Passed to you: I'm tired of carrying this damn bird.' The only sound answer was radio, and as radio slowly improved, so did artillery's use of it, so that forward observers were able to communicate with their guns at any time of day or night without interference from enemy fire or weather conditions.

The problem of predicted fire, born during the 1917–18 battles, was also given some thought.

A manufacturer's section drawing of the Bofors 37-mm anti-tank gun. This was bought by several countries, including the Sudanese Army, and their guns were later used by the British in the Libyan campaign.

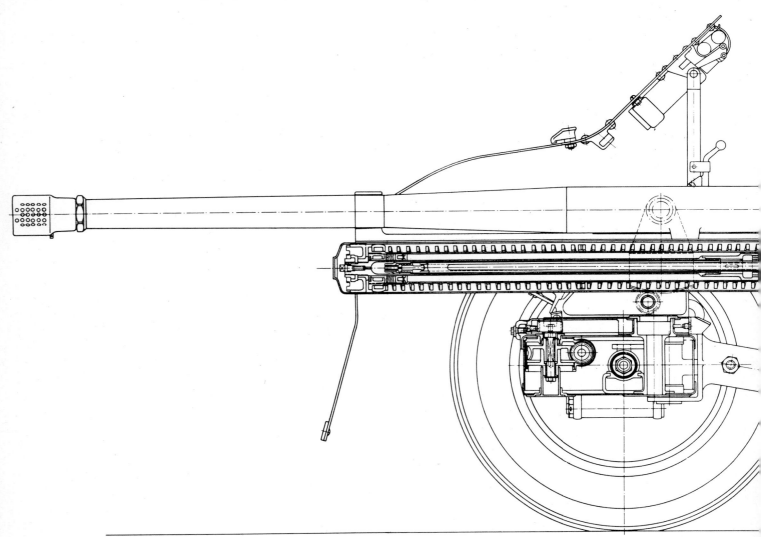

The American 105-mm howitzer M1, the original horse-drawn version which was dropped in favour of the truck-drawn M2 model.

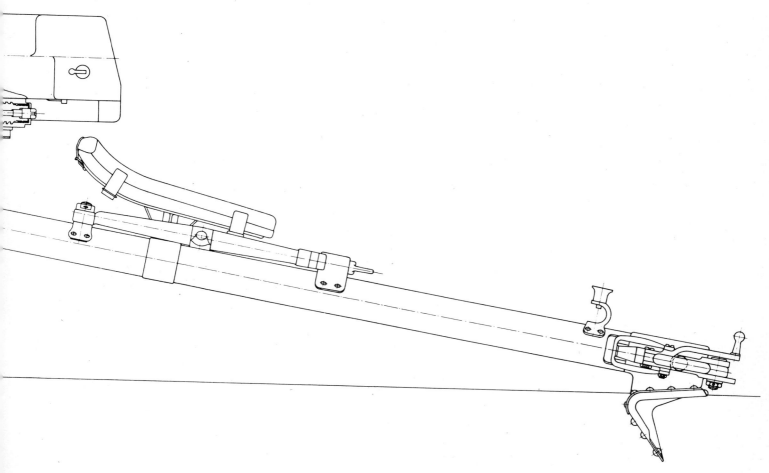

The 45-mm Russian anti-tank gun, Model 1942. Based on a German 37-mm design but of heavier calibre, it was marginally effective and was soon replaced by the 57-mm model.

What was wanted here was the ability to be able to predict the weather and other factors so that fire could be opened upon data taken from a map, without the preliminary ranging which alerted the enemy. Research was done on ballistic questions of what happened to the shell under various conditions of weather, what effect changes in temperature had on the propelling charge, how the wear of the gun affected the velocity and accuracy of shooting, and many other questions, all of which were interrelated. A meteorological service was instituted, usually, as with the British,

drawing its information from the Air Force who were better equipped to provide this information, though the German Army provided *Wetter Pielzug*, sections whose sole object in life was the provision of meteorological data for artillery units.

Finally came the problem of trying to do something about the artillery of the enemy. Flanders had shown the need to find out where the enemy guns were so that they could be attacked prior to any advance, in order to reduce the effectiveness of their defensive fire, and

Another modernized gun is this French 105-mm Model 1913; originally provided with iron-tyred wooden wheels, the axle and wheels were changed during the 1930s to allow high speed movement.

An interesting experiment of 1942 was the fitting of the US 105-mm howitzer into a half-track to become the 'Gun Motor Carriage T19'. A few were used in North Africa but the design was not standardized.

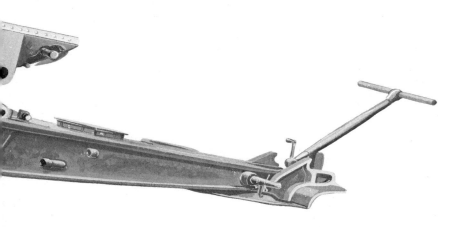

attacked at other times to reduce the volume of their offensive fire on to one's own troops. The simple technique of flash spotting was born in the early days of the First World War; here observers watched for the flashes of enemy guns and, by cross-observation, were able to deduce their locations. This technique suffered a setback when flashless propellants were developed, and it was reinforced by 'sound ranging' in which a number of microphones were buried behind the front line and connected to an instrument which recorded the time of arrival of the noise of a hostile gun. By a piece of elegant geometry it was possible to use this to deduce, with surprising accuracy, where the enemy gun was located. The system had its beginnings in a suggestion by a Second Lieutenant Corry of the Royal Field Artillery in 1914; it was taken up and examined more closely, improved by Sir Lawrence Bragg and put to use. In the post-war years it was improved even more until it became possible not only to find the enemy guns by the sound of their firing but also to shoot back at them, comparing the record of the bursting shells as picked up by the microphones and adjusting the fire until the records of gun sound and shell burst were identical, at which point, obviously, shell burst and gun location coincided.

The guns were built; the systems were proven; the ammunition was ready; the techniques were practised and perfected; the machine was tuned up and ready to go. All it needed was somebody to press the button. In the late summer of 1939 Adolf Hitler looked towards Poland and reached out his hand.

CHAPTER 8

Battle is Resumed

The Second World War does not lend itself to a study of artillery's progression in chronological fashion; the nature of the war, with more accent on movement led to the development of different types of artillery on different lines concurrently, and it is therefore better to look at the various types in isolation, although there was a considerable amount of crossing of lines as ideas developed in one application spun off into another. In any event, the opening phase of the war saw little of remark in artillery employment; the gunners were too busy finding their feet and tidying up the loose ends of technique developed in peacetime, but which tended to become unravelled at the edges when subjected to the test of war. Radio communication, for example, was a fine thing when done by a small peacetime army of professionals with a sufficiency of long-trained signallers and a back-up of handy repairmen, but when the signallers were rusty reservists or hastily trained conscripts and the repairman was 100 miles away and already inundated with work, even the best communication systems developed a stutter.

The anti-tank battle was the first to show signs that all was not well with the guns. In pre-war days the tank had been relatively feeble, and the standard anti-tank gun of most nations was a weapon of about 37 to 40-mm calibre, firing a projectile of about 2 lb weight. The battles of 1940 demonstrated the limitations of these guns: unless the tank was engaged at a very short range the shot might not penetrate; when it did, the damage was minimal. And if the tank saw the gun first, the odds were it would be able to stand off and shoot up the gun before it came within the anti-tank gun's lethal range. Heavier weapons were needed. To be fair, the gunners had appreciated this long before, having watched the rise of the tank and made a forecast of what the tank would become, and they had demanded heavier weapons. Both Britain and Germany had designed for the future in 1938, Britain proposing a 6-pounder 57-mm gun and Germany a 4½-pounder of 50-mm calibre. But in Britain the need for field and anti-aircraft guns was uppermost, and production facilities were not available to produce the new gun. In Germany, too, production was hard at work with other things, and it was not until late in 1940 that their improved gun reached the troops.

With this move as a starting-point the battle for

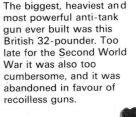

The biggest, heaviest and most powerful anti-tank gun ever built was this British 32-pounder. Too late for the Second World War it was also too cumbersome, and it was abandoned in favour of recoilless guns.

Comparable to the British 32-pounder for power, but lighter and more handy in action, was this German 128-mm gun of which few got built before the war ended.

supremacy between the tank and the gun was fairly joined, and continued at full speed until the war ended. The basic need was always to provide a gun firing a heavier shot at higher velocity to a longer range, and, of course, it followed that the gun gained in size at every turn. Germany followed their 50-mm with a 75-mm, then an 88-mm and finally a 128-mm firing a 62-lb projectile. Britain followed her 6-pounder with a 17-pounder and at the war's end had a 32-pounder undergoing trials. But from an all-up weight of 952 lb for the German 37-mm with which they entered the war to 10 tons of 128-mm gun, the problem of manœuvring the weapon and handling it on the battlefield had got out of hand. Brute force was not enough.

Science had been called in to assist in the earliest days, Germany being the first to approach the anti-tank problem from a different angle. The basic problem was that simply flinging a pointed chunk of steel was no longer sufficient; as coast and naval gunners had found in years gone by, the shot shattered when it struck the target at high velocity. The same solution was applied—to put a cap on the shot to help it to penetrate. As tank armour became thicker and harder, even capped shot began to fail and it was time to find a better solution.

At the turn of the century a German inventor named Karl Puff had suggested that if the barrel of a gun could be made to taper in calibre as it reached the muzzle, then provided a suitable projectile could be constructed which would also decrease in calibre as it passed up the bore, the muzzle velocity would be enhanced. This was due to the base area of the projectile reducing while the propelling gas pressure remained the same; this placed a greater unit pressure on the shot and hence pushed it out more quickly. As well as patenting the idea, it appears possible that Puff actually made one or two experimental small arms to test his theory—the 1904 catalogue of a German cartridge company lists a 9-mm experimental Karl Puff rifle cartridge—but he ran into trouble making a bullet which would shrink, and his work was never taken to a practical conclusion. The idea was later taken up by another German experimenter called Gerlich, who used it to develop high-velocity hunting rifles which were commercially marketed in the late 1920s, and he also attempted to interest various military authorities in their use as sniping weapons, without much success. After doing experimental work for both the British and US armies he returned to Germany in the early 1930s and vanished from sight. But his idea was taken up and turned into a workable anti-tank

A version of the German 128-mm anti-tank gun, this one developed by Krupp.

A 17-pounder 'Straussler' anti-tank gun. This is the standard gun with the addition of an engine and a third wheel, to give it some degree of mobility independently of its tractor. It was not taken into service, largely due to the difficulty of concealing it in the firing position, a prime requirement for anti-tank guns.

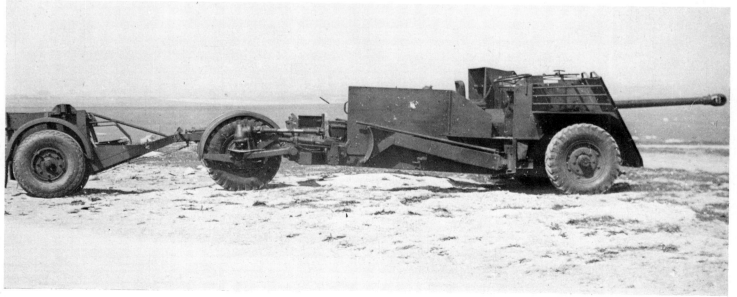

Another wartime idea for making anti-tank guns more mobile was this British 'Prime Mover', a wheeled and motorized framework into which a 6-pounder gun could be run for short moves.

An American design of anti-tank gun was this 90-mm T8. Based on a very good anti-aircraft gun, it was married to a number of carriages but eventually saw service only in self-propelled or tank mountings.

gun. Issued to German units in the Libyan Desert in 1941, the 2·8-cm Schweres Panzerbuchse 41 had a barrel which tapered from 28-mm calibre at the breech to 21-mm at the muzzle; this gave the projectile the phenomenal velocity of 4,590 feet per second and the capability of piercing 66 mm of armour at 500 yards.

Obviously the projectile was far from conventional; an ordinary shot or shell could not be reduced in diameter as it went through the barrel, nor would it have survived impact on the target at such high velocity. The Gerlich shot used a central core of tungsten carbide supported in a soft-iron sheath with skirt-like driving and supporting bands at shoulders and base. These were malleable enough to be squeezed back as the bore contracted, keeping the mass of the shot central and still forming a seal for the propelling gases. By the time the shot reached the muzzle the skirts

had been pressed down until they were almost flush with the body of the shot, which thus presented a streamlined surface for a smooth airflow and good flight ballistics.

The 28-mm model was later followed by 42-mm and 75-mm versions, but these, while efficient, were short-lived. Tungsten carbide, the essential component of their projectiles, was a scarce material, and what supplies Germany could get were insufficient for both ammunition of this kind and for the production of machine tools. In the end it came down to a direct choice between them as to how the available tungsten was to be allocated, and Hitler came down on the side of machine tools. The taper-bore gun was doomed from then on, and as soon as their supply of special ammunition was used up they were withdrawn from service.

Tungsten was not only used for squeeze-bore

Projectiles for squeeze-bore guns; on the left a 2·8-cm in section, showing the core of tungsten carbide, on the right a 75-mm model. The holes in the forward skirt release air which would otherwise be trapped in the squeezing process.

A British 25-pounder in its more usual indirect-fire role.

firstly the economy of material obtained by only making the core of tungsten, and secondly the inescapable physical fact of tungsten's density. It was almost double the density of steel, so that a full-calibre projectile would weigh twice as much as a conventional steel one. Hence the propelling charge would have to be reduced, otherwise the pressure might well blow the gun apart before such a super-heavy projectile began to move, and thus the velocity would be a lot less than with a plain steel shot. But by making the shot in composite form, the result was some degree lighter than an all-steel projectile would have been, hence the charge could be increased and the velocity given a boost, a welcome bonus in the circumstances.

Projectiles on these lines were tried by Britain, Germany and the USA; Germany had to abandon their work in 1942 due to the tungsten famine; Britain and the US continued to study the idea, but Britain soon abandoned it, since they discovered that the system had a defect. The ratio of weight to cross-section was poor and thus the projectile lacked 'carrying power' (for the want of a better term) and tended to lose velocity fairly rapidly, until at 1,000 yards' range its performance was actually rather worse than that of a conventional steel shot. The US persisted in research and eventually produced some remarkably efficient projectiles of this 'composite rigid' variety which were to remain in service for many years. In Britain the designers preferred a 'non-rigid' solution; the first attempt was to provide the standard 2-pounder guns with a squeeze-bore adapter on the muzzle which allowed them to duplicate the performance of the German guns without going

ammunition; with the increase of gun velocities to the shatter-point for steel shot, tungsten was adopted as the solution for parallel-bored conventional guns as well. Here the heavy and hard material was used to form a central core which was then built up to the requisite calibre by light steel or alloy sheathing until it fitted the bore. The reason for this form of construction was twofold:

(Far right) The 25-pounder on tow behind the 'Traclat', an experimental artillery tractor of 1944, copied from the well-known German three-quarter-tracked vehicles.

As well as adopting the French 155-mm howitzer, the US Army took the partner 155-mm GPF (Grand Puissance, Filloux) gun. It remained in service until the Second World War, and was then taken into use as a self-propelled gun, mounting the barrel and cradle on a redundant tank chassis.

to the trouble of manufacturing new barrels. This was reasonably successful, but the users didn't like it because it could only be used with special projectiles, and they had to run round to the front of the gun and unscrew the adapter whenever they wanted to fire conventional ammunition. The eventual solution was the perfection of 'discarding sabot' projectiles.

As with most other ordnance ideas, sabot ammunition is quite old; there are patents dating from the 1870s which cover the general principles, but like many other ideas, the brainwaves of the inventors were a long way ahead of the ability of the engineers, and the idea had to lie dormant for many years. It was revived during the 1930s by Edgar Brandt, a brilliant French ordnance engineer who was never afraid to try an unorthodox solution; the war interrupted his work, but enough was known about it in other countries to allow both British and German designers to use his work as a foundation from which to produce service ammunition.

Brandt's original intention was to give guns longer ranges; a 155-mm gun, for example, could be provided with a 120-mm projectile held inside a 'sabot' or sleeve of 155-mm calibre. The all-up weight would be less than that of a 155-mm shell, so the muzzle velocity would be greater. As the combination shot left the muzzle, so the sabot would fall clear and allow the 120-mm 'sub-projectile', propelled by a 155-mm charge, to depart for the target, and due to the extra charge it would go to a much greater range than either a 120-mm or a 155-mm shell fired from conventional guns. The principle was worked on by the Germans during the war years and they eventually produced a number of sabot shells for several weapons in order to increase their maximum range. The principal was also applied to anti-tank weapons, in order to launch the sub-projectile at a much higher velocity than could otherwise be attained; numbers of sabot projectiles were made in which the sub-projectile was the steel anti-tank shot of the next lower gun calibre; thus an 88-mm gun would fire a 75-mm sub-projectile.

In Britain, where the shortage of tungsten was not particularly acute, the sabot principle was used to propel tungsten carbide sub-projectiles at high velocity, and the resulting penetration was such an advance on previous shot that the sabot was

German 17-cm gun on the 21-cm howitzer carriage, one of the standard guns of the German Army during the Second World War. This design used a dual recoil system in which the barrel recoiled in the cradle and the top carriage recoiled across the platform.

The German 15-cm heavy howitzer Model 18, a standard medium gun which fired a 95-lb shell to 14,500 yards.

taken into use as the standard anti-tank projectile in 1944 and has remained so ever since. In post-war years it was also adopted by the US Army to replace their composite rigid projectiles; but there seems to be a certain amount of art in the design

and construction of a successful sabot shot, which the British designers seem to have mastered. In any event, the NATO standard discarding sabot shot in use today is entirely British design and manufacture, irrespective of the language stencilled on it.

However, the application of tungsten carbide projectiles to anti-tank shooting demanded large and heavy propelling charges to give the shot the desired velocity, and this in turn demanded large and heavy guns, until the upper limit of practical size was reached—and indeed, in one or two designs, surpassed. Moreover, there is no guarantee in war that tanks will obligingly present themselves in front of anti-tank guns; they might equally well appear in front of an ordinary field gun, and therefore it became necessary to provide almost every gun with some form of anti-tank projectile. Once tanks gained in strength, the conventional field gun, with a muzzle velocity in the 1,700 feet per second region, could not hope to propel a steel shot with any hope of piercing, and at such low velocities tungsten was wasted. As a result, it became necessary to find some other method of attack less dependent upon velocity.

Once again, a nineteenth-century idea had reached practical status just in time to be brought into the battle at a critical time. It had long been known that cutting a conical cavity in the face of an explosive charge before placing it in contact with a target would improve the penetration effect of the charge. Many years had elapsed while inventors tried to turn this scientific novelty into a practical weapon, and the breakthrough had come

Sgt Salmon of the Royal Canadian Artillery, sight-testing a British 7·2-inch howitzer. Note the loading tray and rammer on the ground.

The 7·2-inch howitzer fires. The next shell is lying ready on the loading tray.

These photographs show stages in the emplacing of an American 8-inch gun, here being operated by British troops. The mounting is lifted from its transport waggon by crane (*right*) and lowered over the prepared pit. The barrel is then brought up and lifted by crane (*below*) and lowered on to the mounting where it is secured by massive bolts (*opposite, top*). Finally the shell is rammed (*opposite, below*) and the gun made ready to fire.

during the 1930s. It was found that the fundamental requirements were that the cavity should be lined with some dense metal – copper was found to be ideal – and that the charge performed best when detonated some small distance away from the target. When this was done, the detonation of the explosive converted the metal liner into a fast-moving jet of metal particles and explosive gas which was capable of pushing its way through any

armour. By degrees, workable projectiles were built around this 'hollow charge' system, beginning with simple grenades and finally reaching artillery shells. Since the effect was dependent upon the explosive substance and the geometry of the cavity and its lining, it mattered not whether the shell was stationary or travelling at 1,000 or 2,000 feet per second when it detonated, so that low-velocity guns and howitzers could now be provided with an anti-tank shell with some guarantee of success.

In the anti-aircraft war the problem was not one of penetrating: it was one of hitting the target at all. The guns which were thought adequate in 1936 were rapidly made obsolescent by the surprising rate of improvement in aircraft performance, both in speed and height of operation. When the British 3·7-inch anti-aircraft gun was first proposed in 1928 it was thought that if it fired a 25-lb shell to 28,000 feet, this would do very well; by the time it entered service in 1937 the specification had changed to a 28-lb shell and 32,000 feet, and before the war was very old it became apparent that something better would have to be provided. A proposed 4·7-inch design had never managed to reach perfection, and a number of 4·5-inch guns were obtained from the Royal Navy in 1936 to provide heavy protection for dockyards and naval bases; this gun sent a 54-lb shell up to 34,000 feet but was only capable of firing eight rounds a

A Canadian lightweight version of the 40-mm Bofors light anti-aircraft gun, primarily intended for airborne use.

An American 75-mm pack howitzer in use by British troops in a direct fire role. This weapon was extensively used by airborne troops.

minute, and since the speed of the bomber was apparently rising daily, a faster rate was required. In the end, after studying several suggestions, a design reminiscent of the First World War 13-pounder was accepted: the 4·5-inch barrel was linered down to 3·7 inches so that the 3·7-inch shell could be propelled by the 4·5-inch cartridge. With the aid of a mechanical loading device, this sent the 28-lb shell to 45,000 feet at a rate of 19 rounds a minute, and restored the anti-aircraft gun to something like parity with the attackers.

Development took similar lines in other countries; the United States entered the war with a venerable 3-inch model which dated from 1917, and, indeed, was based on the barrel of an even older coast defence gun. This could send a 13-lb shell to about 28,000 feet at a rate of about 20

rounds a minute, but it took a highly trained detachment to do it. A 90-mm gun was introduced just before the war which had a 23-lb shell and a ceiling of 32,000 feet, and this was backed up by a 105-mm gun firing a 33-lb shell to 37,000 feet and a 120-mm firing a 50-lb shell to 47,000 feet. The two latter were little in evidence during the war however; only 14 105s were ever made, the majority of which were installed in the Panama Canal defences, while the 120-mm was considered to be too heavy and powerful for field army use and was retained for the defence of the USA.

In Germany the progression was similar, from 88-mm through 105-mm to 128-mm, and in addition there were proposals to build guns of 15-cm, 20-cm and even 24-cm calibre, firing shells of 435 lb up to 60,000 feet. Such massive weapons were

A British 3·7-inch anti-aircraft gun about to leave the factory. This picture well illustrates the growth in complexity of the gun and its mounting.

A 25-pounder battery goes into action in North Korea, 1952. Pits had been dug in advance; the Number One of one detachment can be seen jumping from his truck in order to run ahead and guide the driver to the exact location of the gun.

to be turret mounted and provided with mechanical handling for the ammunition in order to get a worthwhile rate of fire, but the designs brought so many problems in their train that the whole series of projects was abandoned late in 1943 and the designers put to work on a new idea, a rocket controlled by radio. They may have thought the whole idea a trifle ridiculous, but within ten years the heavy anti-aircraft gun was being pushed from the stage by the descendants of those original German designs.

There were, of course, other aspects of anti-aircraft gunnery; it would be impossible to contemplate anti-aircraft in the Second World War without mentioning the ubiquitous Bofors 40-mm gun, for example. This was first placed on the market by the Swedish company in 1929. During the early 1930s the British Army was casting about

This US 155-mm 'Long Tom' gun is at its maximum elevation of 63° 20'. The carriage was interchangeable between the 155-mm gun, the US 8-inch howitzer and the British 7·2-inch howitzer.

The French 75-mm Model 1897 remained in service until 1945; this shows it in pre-1939 form, with the addition of pneumatic tyres for motorized towing.

The 25-pounder, primarily a field weapon, was also a potent anti-tank gun.

A French two-gun turret mounting 34-cm weapons, emplaced by the Germans to protect Toulon. The piggy-back gun is a 75-mm used for close defence and also as a training weapon to save wear and tear on the main armament.

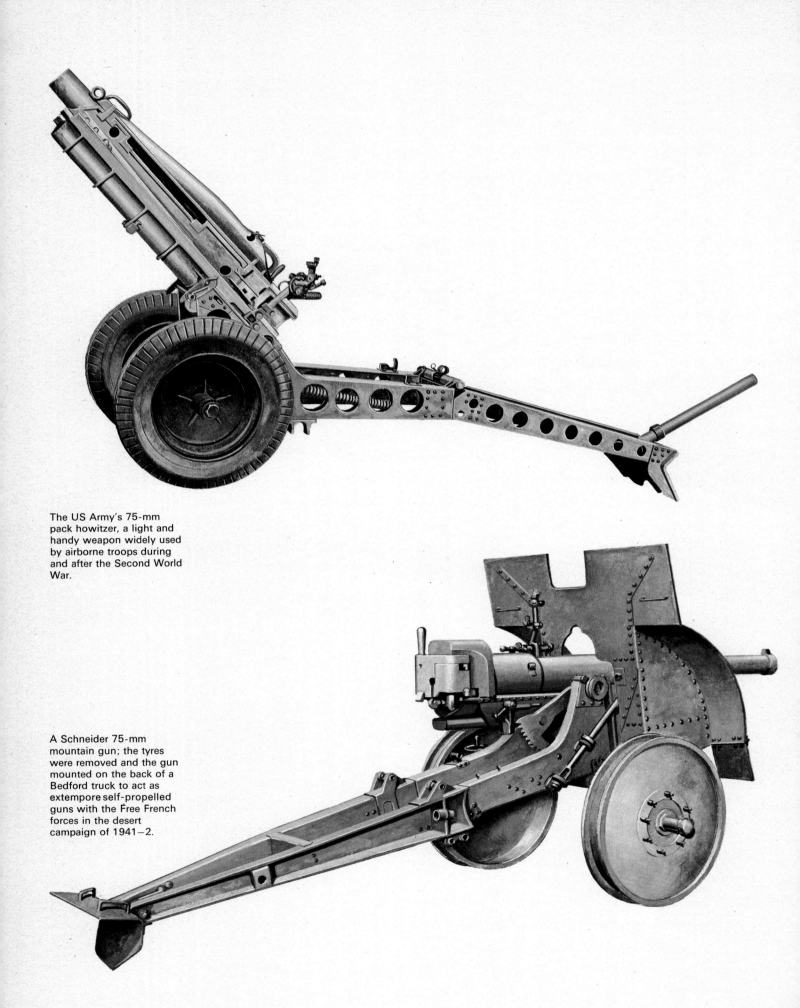

The US Army's 75-mm
pack howitzer, a light and
handy weapon widely used
by airborne troops during
and after the Second World
War.

A Schneider 75-mm
mountain gun; the tyres
were removed and the gun
mounted on the back of a
Bedford truck to act as
extempore self-propelled
guns with the Free French
forces in the desert
campaign of 1941–2.

A Canadian development of the Second World War was this 3·7-inch anti-aircraft gun on a self-propelled chassis. It did not enter service.

The US 240-mm howitzer; this pattern replaced that shown in the earlier picture. The cylindrical tank at the front is a pressure tank for the equilibrators which balance the weight of the gun.

for a suitable lightweight fast-firing gun to accompany troops in the field and, after considering a number of possibilities, adopted the Bofors in 1935. Firing a 2-lb shell to 20,000 feet at a rate of 120 rounds per minute, the Bofors was a highly effective weapon and was soon adopted by almost every nation at war; the one notable exception was Germany. Although a small number, largely captured, were in service as the '4-cm Flak 28', the Germans preferred their own design of 37-mm gun.

Although the Bofors gun could send a projectile up to 20,000 feet it was not effective at this height, for a number of reasons. In the first place much of the Bofors shooting was done by the unaided human eye, and it takes a hawk-like vision to determine the effect of shells four miles up in the sky. But principally the restriction was due to the ammunition; this brings in something which, the reader may recall, was dismissed rather lightly in 1910–the danger of the shells coming back down to earth after missing the target. In order to guard against this, all Bofors shells had a 'self-destroying' device built in which burst the shell in the air should it have failed to strike a target in its flight. This occurred at seven seconds after leaving the muzzle, which restricted the maximum ceiling to 7,200 feet.

A by-product of this was that there was a belt of sky relatively undefended. The light weapons such as the Bofors and the German 37-mm were limited by their self-destruction to about 7,000 feet, while the heavy weapons did not come into their own until about 15,000 feet. Below this they could shoot, but the angular rate of movement of a target was too fast for their traverse speed, and they could not deal effectively with low-flying aircraft. In view of this, efforts were made to close the gap by the development of what became known as 'intermediate AA guns'. The Germans began by designing and building a 50-mm automatic gun firing a 5-lb shell to 18,000 feet; it was a serviceable enough gun, but the mounting was less so, prone to overturn if towed in too exuberant a manner and with a tendency to be unstable when fired. Nevertheless it showed that such a weapon could be useful, and a better 5·5-cm gun was designed. The work was complicated by the fact that the war had by now demonstrated that a single bomber could wreak havoc out of all proportion to its size were it allowed to get through to some types of target, a premise given a new edge by the successful raid on the Ruhr Dams. As a result, the High Command demanded a gun backed up by a sophisticated fire-control system which would unerringly guarantee a 100 per cent kill rate, and if this could be produced they were willing to foot any size of bill to get it. Consequently the '5·5-cm Flak Equipment 58' programme did not stop at the gun design, it encompassed radar, fire control, predictor, remote power control for the guns, and several other refinements, all of which were so long in being brought to a workable state that the war ended before the system was working.

The British Army also appreciated the gap before the war, but, as with many other desirable things, nothing could be done which might jeopardize the production of more basic weapons. Some possible solutions using existing weapons were examined, and it was decided to take the existing 6-pounder as used by the coast artillery and see if it could be turned into an anti-aircraft gun. As a gun there was nothing against it; the 6-lb shell could be sent up to 21,000 feet quite easily, which was enough for the purpose. But providing automatic loading for a gun which had been designed as a hand-loader turned out to involve some frightening mechanical propositions; the entire war was spent on the automatic loading problem without achieving much success. Britain never did get an intermediate gun into service.

After the collapse of France in 1940 the Germans moved their two 21-cm K12 railway guns down to the Channel coast and opened fire against England; the residents of north Kent were somewhat aggrieved to be thus brought under direct shellfire from the French coast. One shell landed

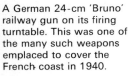

A German 24-cm 'Bruno' railway gun on its firing turntable. This was one of the many such weapons emplaced to cover the French coast in 1940.

Preparing ammunition for the British 9·2-inch howitzer. The detachment commander in the rear is aligning the rammer with the bore, preparatory to giving the order to ram the shell home.

Hoisting ammunition up to the operating deck of the German 24-cm railway gun.

as far inland as Rainham, near Chatham, 55 miles from the nearest point on the French coast. This appears to have been the only combat application of these expensive and complicated long-range guns, but the potentialities were appreciated and the question raised of whether or not similar performance could be obtained at rather less expense in both material and complication. Another railway gun was being produced in some numbers for the Army, the 28-cm Kanone 5 (E) (again, the nomenclature came from the estimated performance, 5 for a range of 50 km). This had a similar type of deep-grooved barrel and fired ribbed shells to a range of 62 km, and in order to reach out further a rocket-assisted shell was designed. This used the forward section of the shell to carry a solid-fuel rocket motor, exhausting through a vent in the bottom of the shell; the blast passed down a central pipe, around which the high-explosive bursting charge was packed, an arrangement which posed a pretty problem in insulation. Ignited by a time fuze after 19 seconds of flight, the rocket delivered an extra thrust while the shell was still on the upward portion of its trajectory and increased the maximum range to 86·5 km, though with some loss of accuracy.

By this time the celebrated Peenemunde Rocket Research Establishment was functioning, and their wind tunnels were available for research into the flight characteristics of a variety of missiles and projectiles. As a result of this, work on a streamlined 'Peenemunde Arrow Shell' was begun, to be fired from a smooth-bore 31-cm barrel mounted in place of the 28-cm rifled barrel on the

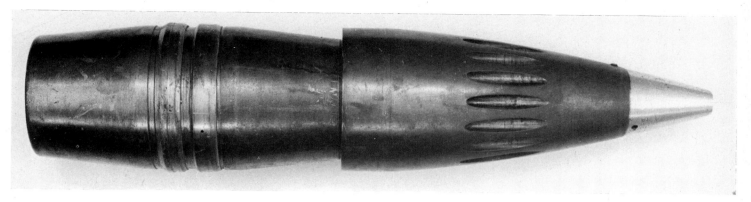

Another German development, a rocket-assisted shell.

(*Opposite*) 'Gustav', the German 80-cm railway gun, prepares to fire a 7-ton shell into besieged Sebastopol in 1942.

Big guns frequently have to travel in pieces; this is the barrel of the German 21-cm Kanone Model 1938 on its transport limbers.

Kanone 5. This shell was a long, dart-like, finned projectile, carrying a centering band about its mid-section which also acted as a gas seal. Fired with an extra-powerful charge it left the gun at 5,000 feet per second, discarded its centering band, and continued to a maximum range of 151 km–93.8 miles. Thus the work-horse railway gun–there were 28 in service by the end of the war–finally gave better performance than the prima-donna K12 long-range gun. Unfortunately for the German Army the development of the Peenemunde arrow shell was not completed until late in 1944 and very few were produced in time to be used; there is only one record of its use in combat, when it was used to shell the US Third Army at a range of over 65 miles.

The Peenemunde arrow shell held out promise of other things beside range; the increase in muzzle velocity, due to the smooth bore and heavy charge, was an asset which could find a use elsewhere, and a 105-mm arrow shell was produced for a smoothbore version of the standard 105-mm anti-aircraft gun. By increasing the velocity the time of flight to the target was reduced and thus the chance of hitting was improved; but not only

was the development late in being perfected, the design also demanded high-grade steel to make the projectile strong enough to stand up to the violent acceleration, and since high-grade steel was in short supply, the arrow shell's promise was never realized in practice.

The arrival of long-range shells in England led to the inevitable demand for something of equal performance with which to shoot back, but, wisely, this was resisted; development of such a weapon would have involved time and effort which could be better employed in other activities. Then, in 1942, when things were a little easier, the Director of Naval Ordnance decided to produce a weapon which would be primarily a ballistic research tool but which might, secondarily, give the occupiers of France something to keep them busy. As a result a naval 13.5-inch gun was fitted with a long 8-inch liner, in much the same sort of configuration as the Paris Gun of 1917. It was installed on a modified 13.5-inch barbette mounting, first on the Isle of Grain and then near Dover, firing northwards so that the fall of the shells could be observed from the experimental range at Shoeburyness. It was hoped that it could then be

turned against German positions in northern France, but the initial trials showed that after no more than 28 shots the barrel liner was worn to the point where the engagement of the ribbed shell into the rifling grooves became critical, and after completion of the trials the weapon was dismantled. In spite of its failure as a war weapon, it produced some useful ballistic information, achieving a muzzle velocity of 4,500 feet per second and a range of about 57 miles.

The 13·5/8-inch 'Hypervelocity Gun' (as it was officially known) appears to have been the only venture into the unconventional by the Allies during the war. The same cannot be said for the Germans; any proposal, no matter how harebrained, had a chance of being accepted, and some weird and wonderful things came to light when the war was over. There was, for example, the device discovered at the Hillersleben Experimental Range by the first Allied troops to reach it; their initial report read: 'At one end of the front was found equipment which cannot be identified. It consisted of an 8-foot cylinder with nozzles leading to the rear . . . made of ¼-inch boiler plate . . . Beside it was an L-shaped cylinder approximately 50 feet in length . . . constructed of iron plate . . . open at one end and pointed down the range.' Subsequent interrogation revealed that this was a 'Whirlwind' or 'Vortex' cannon, a project begun by a Dr Zippermeyer. A mixture of oxygen and hydrogen was burned and the nozzle discharged a whirling vortex of air which, it was claimed, could break a four-inch wooden spar at 200 yards range and, it was intended, would severely damage an aircraft or at least put up such a disturbance in the sky as to cause a pilot to lose

(*Left*) The remarkable 'Smith Gun', a 3-inch smoothbore built for Britain's Home Guard in 1941. Designed by a toy factory and built by a piano company, nevertheless it worked. In action, as here, it sits on one wheel and uses the other as a roof, allowing all-round traverse.

(*Right*) A naval twin 2-pounder anti-aircraft gun. Numbers of these were also emplaced around British dockyards.

(*Below left*) The carriage portion of the German 15-cm Kanone Model 18, with a two-wheeled transport limber fitted under the trail ends.

control of his aircraft. Were one to be attacked by four-inch wooden spars at a range of 200 yards, it would doubtless have been an ideal weapon, but getting the vortex up to a more practical height was an insuperable problem, and the Vortex gun never achieved success.

Another way-out weapon was the 'Sound Cannon' in which a methane and oxygen mixture was detonated to give a sharp bang; this was amplified and sent into the sky by a parabolic reflector. The detonations were rapidly repeated and the effect was to build up a continuous high-pitched tone which, experiments showed, could be lethal at short ranges and extremely painful at ranges up to 300 yards. Again, the short range was never improved upon and the weapon never got any further.

Electricity as a motive power for projectiles has always attracted inventors, and a number of proposals for using launchers based on the principle of the electromagnetic solenoid appeared from time to time. While the solenoid can undoubtedly launch projectiles, the amount of electricity needed to achieve a worthwhile velocity and range, sufficient to take the device out of the realm of a laboratory stunt and into the field, has always proved the stumbling-block. During the First World War a Frenchman came up with the first

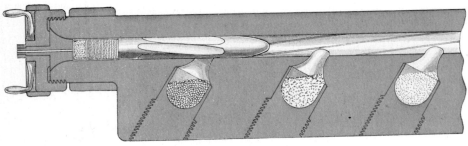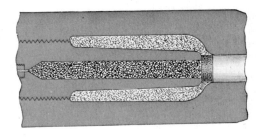

improvement when he designed a winged projectile which spanned two electric conductors, to make the original linear motor. While he was undertaking experiments for the French Government the war ended and the idea was abandoned, but it was revived in Germany in 1943 and the proposal was put across so plausibly that the Luftwaffe were persuaded to give a contract for a 4-cm anti-aircraft gun. Some basic research was done and laboratory models persuaded to work, but again the war ended before it could be brought to perfection. Afterwards it was examined more closely and finally abandoned as a practical proposition when it was shown that each gun would have required a power station all to itself in order to produce the necessary amount of current to make it work.

Almost as visionary, in the eyes of conventional ordnance engineers, was the proposal of one Herr Conders to produce a multiple-chamber gun with which to bombard England from the area of Calais. The idea of the multiple-chambered gun was far from new: as the 'Lyman and Haskell Gun' it was tried out in the United States in the 1880s – and failed miserably. The basic idea is to build a long-barrelled gun with auxiliary side chambers, each containing a cartridge; the shell is fired by a cartridge in the normal way, and as it

passes each side chamber the charge therein is fired, so as to add to the gas pressure behind the shell and thus increase the velocity. As can be seen from the drawing of the Lyman gun, the Whitworth system of rifling was used, and when it was fired the inventors were distressed to discover that it actually achieved less velocity than a conventional Whitworth gun of the same calibre. Investigation showed that this was due to the ignition flame passing over the shell and igniting the side charges in advance of the projectile's passage, thus acting in opposition to the shell instead of assisting it.

Herr Conders felt that it might be possible to improve on this with modern techniques, and built a 20-mm prototype which performed more or less as forecast. With this behind him he persuaded Hitler to approve the building of a 15-cm-calibre gun of 50 barrels, built into a hillside and pointed at London. Work duly began, before the actual construction of a working gun in 15-cm calibre had been managed. Numerous problems bedevilled the project, and the Allies captured the firing site before it could be got into action, which was probably just as well. Odd as Conders's idea may sound, he was within a short distance of making it completely successful. Two shortened versions were built and used against the

An elderly Italian 75-mm field gun, a typical design of the 1920s.

The German light field howitzer Model 18/40, an attempt to reduce the weight of the standard field gun by assembling it to the carriage of the 75-mm anti-tank gun. It saved only 65 lb.

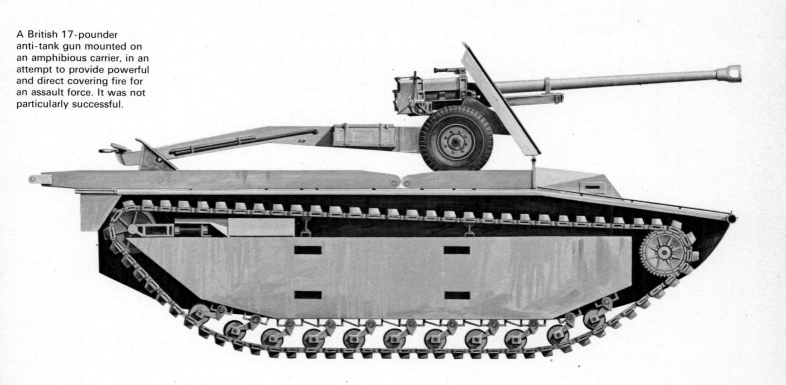

A British 17-pounder anti-tank gun mounted on an amphibious carrier, in an attempt to provide powerful and direct covering fire for an assault force. It was not particularly successful.

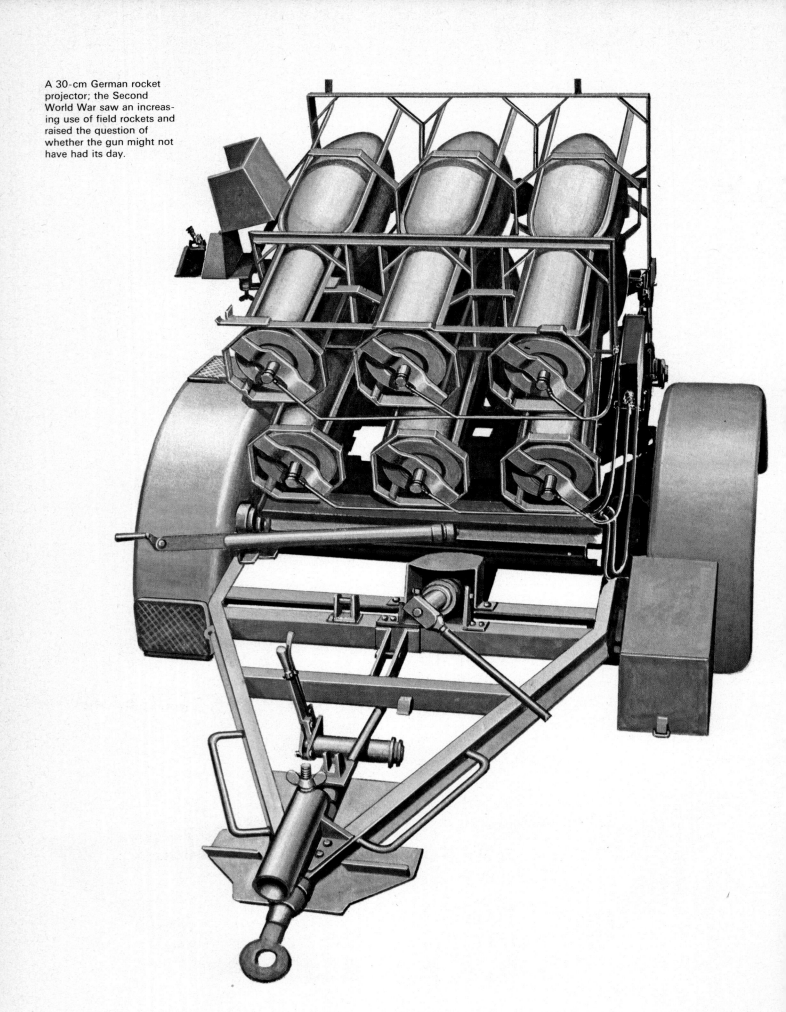

A 30-cm German rocket projector; the Second World War saw an increasing use of field rockets and raised the question of whether the gun might not have had its day.

A British 5·5-inch medium gun being used in the direct-fire role, not a very common application but one which could be very effective.

The 5·5-inch being loaded. The shell is being rammed, while in the foreground the cartridge is about to be carried forward to be loaded. The gun layer is setting his sights, while the breech operator re-loads the firing lock.

Allies in the neighbourhood of Luxembourg, but they were blown up before they could be captured, and there is no record of their effectiveness in action.

More conventional in their mechanics, but scarcely credible in their calibre, were the super-heavy weapons ranging from the 54- and 60-cm self-propelled howitzers known as the 'Karl Equipments' to the legendary 80-cm 'Gustav' railway gun which was used to shell Sebastopol in 1942. These were simply normal guns carried to extreme dimensions. Gustav, for example, weighed 1,350 tons when assembled for action—a task which took the better part of six weeks—and fired a 7-ton shell, while the 60-cm 'Karl' could be motored into position and could fling a 2-ton shell to 4,900 yards. Not a very impressive range perhaps, but sufficient for the task at hand, since 'Karl' was simply the modern successor to 'Big Bertha', a specialized weapon designed for the purpose of attacking fortifications.

The self-propelled mounting advanced in leaps and bounds during the course of the war, but it was noticeable that there were two different schools of thought on the subject, represented by the British and Americans on the one hand and the Germans and Russians on the other. Generally speaking the British and American object was simply to put a standard field piece on tracks so that it could keep up with the armoured divisions and act in its normal indirect-fire support role, driving into position and then performing in exactly the same way as a towed gun would have done. On the German and Russian side, their approach was to develop short-range large-calibre weapons to act as direct support for the infantry during their advance, the guns driving forward like tanks and engaging strongpoints over open sights as opportunity offered. True, the divergent opinions spilled over into each other's camp; thus the British produced a 95-mm tank howitzer which was little else than a close-support gun, while the Germans put their standard 105-mm field howitzer on to a variety of chassis to act as indirect-fire weapons, but these can be considered the exceptions which prove the rule. By far the majority of Russian and German development was in the assault-gun field, while most of the British and American weapons were purely indirect-fire with no ability to perform as assault guns.

British development of a self-propelled gun was held back by the lack of a suitable chassis or the manufacturing capacity for developing and pro-

ducing a purpose-built one; the first serious attempt was made by putting the 25-pounder field gun into a large square box on top of a Valentine tank chassis. It worked, and the crews did their best, but the arrangement prevented the gun's full elevation being applied and thus the maximum range of the standard field gun was not forthcoming. The first really effective self-propelled gun in the Allied camp was the American M7, in which their standard 105-mm field howitzer was married to a standard tank chassis; with an open top, the howitzer was only capable of 35 degrees of elevation, which again restricted its range, but it was a serviceable weapon and pointed the way that design should go. The Canadians now produced a tank called the 'Ram' loosely based on the American Sherman, but by the time it was ready for production it was obsolescent and outgunned, and someone had the happy inspiration to remove the turret, build up the hull to form an open-topped superstructure, and mount the 25-pounder gun into it to produce what became known as the 'Sexton', a thoroughly efficient weapon which allowed the gun to reach its full range and which remained the British service self-propelled gun until the 1950s.

The urge to go bigger is always about, and the Americans were unable to resist it in the self-propelled field; they eventually took their 240-mm howitzer and placed it on a chassis to produce a formidable weapon intended for use against the obstacles and defence works anticipated on the mainland of Japan. A small number were made and readied for shipment, but the end of the war

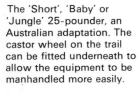

The 'Short', 'Baby' or 'Jungle' 25-pounder, an Australian adaptation. The castor wheel on the trail can be fitted underneath to allow the equipment to be manhandled more easily.

(*Above*) Japanese Model 94 75-mm mountain artillery gun, designed to be rapidly stripped down into eight mule-pack loads.

(*Right*) A German projectile for the longest-range gun of all, the 15-cm Conders 'High Pressure Pump', a multiple chambered gun built into a hillside and aimed at London.

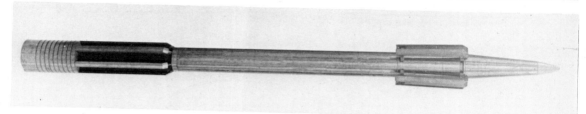

The final version of the French 75-mm Model 1897 was this 1930s version with pneumatic tyres. Otherwise it was little changed from the original.

put an end to their purpose and after a few years they were scrapped.

The credit for the largest-calibre gun of the Second World War, and indeed, together with the Tsar Puschka, the largest-calibre gun of all time, goes to the US Army for their 'Little David' rifled muzzle-loading howitzer. This was of 36-inch calibre and fired a 3,650-lb shell to about six miles range. The weapon was originally developed as a

proof device; when testing aerial bombs against armour or concrete it is difficult to guarantee hitting the target from something like 25,000 feet altitude, particularly when the target happens to be a six-foot square of experimental plate or something equally small. But is quite in order to launch the same bomb from close range by means of a gun, using a powder charge which will deliver the bomb on to the target at the same

The 'Triple Polsten', three 20-mm cannon on a motor chassis, which provided anti-aircraft protection to British vehicle convoys.

A French design of more modern aspect was this 47-mm Puteaux anti-tank gun, using a three-legged mounting to obtain all-round traverse.

The German light field howitzer 18, standard divisional gun of the German Army throughout the Second World War.

velocity as it would achieve if dropped from operational height. Most nations use redundant large-calibre howitzers for this task; but the Americans decided to build a really big one with which they could launch all manner of bombs of the really large varieties, and during the development, some unrecorded hero was inspired to turn it into a siege gun, again for the attack of Japanese defences. As with the 240-mm SP howitzer, the war ended before it could be got into action.

While on the land fronts the gun was assuming greater and greater importance, bigger and bigger calibres vying for notice with enormous numbers of lesser guns – the Russians deployed almost unbelievable numbers of guns towards the end of the war, over 32,000 being used in an attack on the Lower Vistula in 1945 – the sea war saw a

reduction on the importance of the gun in some respects. The first major sea action was the fight of the *Graf Spee*, and while this was a British victory, it was a close-run thing in many respects, not least of which was the poor showing made by the British armour-piercing shells. It looked very much as if little or nothing had been learned since 1918 in this field and urgent inquiries were put in hand to see whether or not the projectiles could be improved. This led, eventually, to the setting up of a Government Ordnance Factory to manufacture piercing shells instead of relying upon contract manufacture, since it was very difficult to assess the worth of a shell on trial unless every detail of its manufacture was known and the contractors were reluctant to disclose some of their secret processes. In fact there was little call for

The Americans also went into the field of super-heavy self-propelled guns with 'King Kong', a 240-mm howitzer on a modified tank chassis. Few were made and none ever saw action.

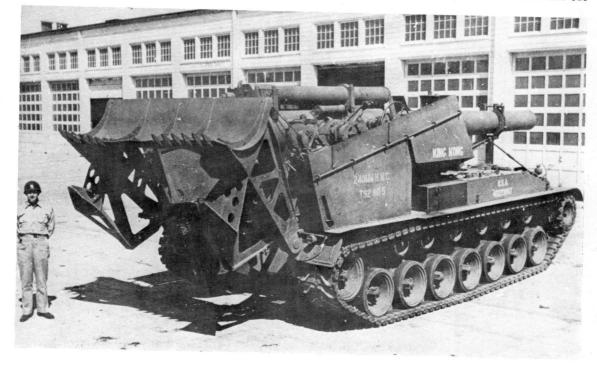

'Little David', the US Army's 36-inch muzzle-loading howitzer of 1945. Only one was built, and the war was over before it could be shipped to the front.

(*Opposite*) 'Little David' takes the road. The muzzle rests on an eight-wheeled dolly, while the elevating arc forms the connection to the M26 Tractor, which normally towed a tank transporter.

Typical of German research during the war is this *minengeschoss* or high-capacity shell; due to its length it was stabilized in flight by four spring-out fins, the driving band acting only to seal the propellant gas.

piercing shells after that, since there was almost no ship-to-ship actions in which piercing shell played a vital part. The aeroplane, carrying torpedo or bomb, began to take the place of long-range naval artillery, and naval gunnery became concerned with this new threat. From this stemmed one of the war's greatest inventions.

When shooting at an aircraft the first problem is to get the shell to within lethal distance of the target so that when it bursts it will do some damage. The second problem is to actually burst it there. Since the first days of anti-aircraft fire this was done by a time fuze, either a powder-burning type or, in the Second World War, a mechanical

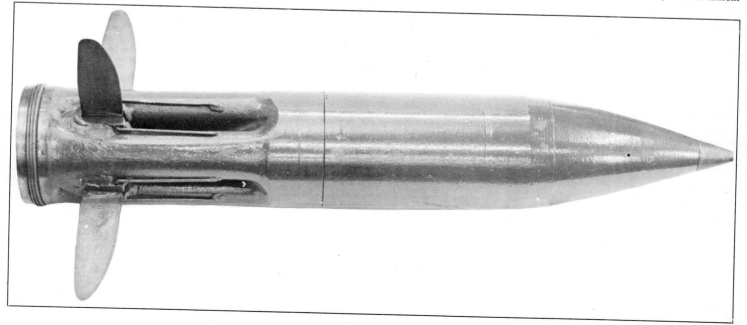

type depending on the running of a clockwork mechanism. The range to the aircraft having been estimated, the fuze was then set and the shell fired; at the set time the fuze would detonate the shell. The use of time fuzes introduced several possible sources of error: the range might be wrongly assessed, leading to the wrong time being set; the time spent in setting the fuze was time during which the target was moving, and this had to be allowed for in calculating the correct setting; there might also be human or mechanical error in the actual setting; and when all this was done there was a considerable possibility that the mechanism of the fuze would be slightly inaccurate. With a shell travelling at 2,000 feet per second an error in any of these to the extent of a tenth of a second could mean missing the correct burst point by 200 feet, and one-tenth of a second was a very small error indeed.

What was wanted was a fuze in which the presence of the target would somehow be detected and the fuze detonated at the correct lethal distance irrespective of human error, a requirement which on the face of it sounded impossible. An attempt was made in the 1930s by a Swedish engineer who proposed placing a strong light source in the base of the shell so as to shine sideways. As the shell passed the target the light would be reflected back, strike a photo-electric cell, and thus generate a current which could be used to trigger the fuze. This was theoretically possible, though it was unlikely that such a delicate arrangement would survive firing from a gun, and besides the basic premise was unsound – the shell would detonate past the target, instead

When the war was over many strange guns were found in German experimental establishments; one of them was this 10-cm anti-tank gun which never got into service.

A German 'tank destroyer' the 88-mm gun mounted on a modified tank chassis, known originally as 'Hornet' and later as 'Rhinoceros'.

of below it, and thus most of the fragmentation would be wasted.

But the germ was there, and the idea was pursued in Britain during the following years, since at that time much work was being done in Britain on developing free-flight rockets as anti-aircraft weapons, and since a rocket accelerated comparatively slowly and gently, there was a chance that such a device might survive the launch and function at the target. Eventually a photo-electric fuze was developed and issued late in 1940; its circuitry was 'tuned' to the normal intensity of light, and when the rocket passed close to an aircraft so that the aircraft's shadow fell across a sensitive cell, the fuze detonated the high-explosive warhead. In spite of being somewhat sensitive—it reacted to passing birds and cloud shadow—it was a successful fuze, but its usefulness came to an end when the German Luftwaffe began raiding Britain at night.

By this time, radar was an accepted weapon of war, and some of the early radar experimenters began to wonder if it might not be possible to put radar receivers into the shell fuze so that they could pick up the reflections from the target as it approached. Trials soon showed there was little hope of making a workable fuze on these lines, and thoughts turned to the prospect of putting a complete radar transmitter and receiver inside the fuze—and this before the days of transistors. Designs were drawn up in 1941 but since there appeared to be no prospect of finding the necessary manufacturing facilities, the whole idea was presented to the United States Government by the Tizard Mission. The US Navy, contemplating the Japanese air strength, saw the advantages of the fuze and took on the task of turning the ideas into a workable weapon. Aided by US industry they were eventually successful, and the first 'Proximity Fuze' was fired in action by the USS *Helena* in the summer of 1943. After this it was used in increasing numbers in naval AA fire, and one spokesman for the US Navy is reputed to have said that without the proximity fuze the war in the Pacific would have been vastly more costly for the Americans.

The proximity fuze then found its way to land force applications being used extensively and with good effect against the V-1 pilotless bombs which were launched against Britain in 1944. Then, on 18 December 1944, they were first used in ground combat by the US First Army, a special model having been perfected which reflected its signal from the ground so as to burst a high-explosive shell at lethal height above the ground for the attack on troops in the open.

CHAPTER 9

Decline but not Fall

The Second World War came to an end in a haze of atomic dust, but some of the argument which raged around the prospects of atomic warfare were a good deal more insidious and potentially lethal than the bomb itself. For several years after the war 'the bomb' conditioned everybody's thinking much as, one supposes, the advent of gunpowder had caused the knights of old to scratch their heads and argue. Air power had also burgeoned during the war years, and bombers were carrying loads unthought-of five years previously; and to crown it all those ingenious Germans had actually invented a rocket which could be guided to its target and which had a range in excess of any gun. Admittedly it was primitive, and the target had to be the size of a large city–but so was the target for the Paris Gun, come to think of it. There was also the fact that the rocket needed very little in the way of a launching device, it was well-nigh undetectable and it arrived at the target with supersonic speed to deliver no less than a ton of high explosive with devastating effect. The other craft which war had incubated was that of electronics, and the postwar years saw much energy and money put into attempts to improve on the German missiles, develop new guidance systems and methods of control, and, inevitably, the atomic scientists were called upon to stuff one of their potent devices into the result.

'Bumble Bee', a German self-propelled 15-cm howitzer; a general arrangement drawing from the original vehicle handbook.

The end of that particular story is not yet in sight; every day we hear of new missiles, anti-missiles and anti-missile-missiles either being introduced or being scrapped. All this furious activity was not without its effect on artillery. The first casualty was coast defence; it no longer made sense to install powerful guns around a dockyard when a missile could be launched from another continent which would demolish the entire base in one mushroom-shaped bang. In 1950 the United States disbanded their coast artillery and in 1956 the British followed suit; many smaller nations, however, who feel that whatever threat they have to face will come from someone as small as them and not from a missile-wielding power, still retain coast artillery, often having equipped themselves with the latest British and American equipment at clearance prices.

Next on the list for disbandment was anti-aircraft artillery, but in this case it took rather longer. The development of serviceable missiles took much more time than was originally estimated, and therefore there was still time for a fresh generation of anti-aircraft weapons to be developed. In Britain the demand was for a fast-firing heavy gun capable of reaching up and hitting intercontinental bombers, while the American Army, well provided with 120-mm guns, was more concerned with providing the field army with a fast-firing mobile weapon.

The British had ended the war with their 3·7-inch Mark 6 guns, the lined-down 4·5-inch weapon. This was powerful and effective and its ballistics were enough to guarantee it a place in the defences for some years; the only argument

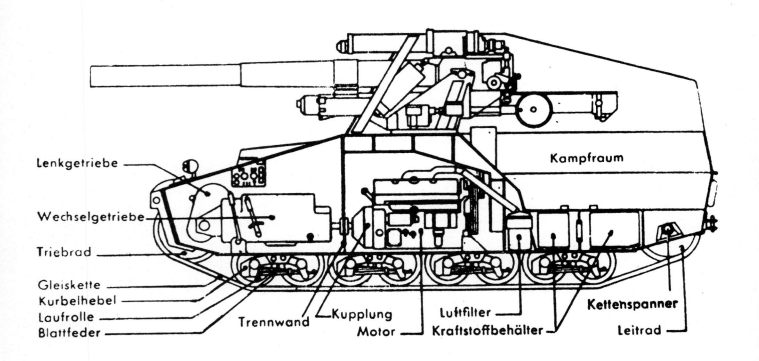

Another German 15-cm self-propelled gun, this time known as 'Grizzly Bear'.

An interesting development by Krupp of Essen; the 'Grasshopper' which trundled into position as a self-propelled mounting, then removed its turret and dropped it into place to act as an armoured pill-box, after which the chassis could be driven off to act as an ammunition supply vehicle.

against it was its rate of fire, and to try and improve this a development programme called 'Rate-fixer' was begun. After studying high-speed film of guns being loaded and fired, a variety of systems of rapid loading were developed, all applied to the 3·7-inch Mark 6 gun, which culminated in a weapon called 'Longhand', a 3·7-inch with rapid loading by a conveyor-belt, firing at the astound-

ing rate of 88 rounds per minute. Meanwhile work had begun on a new gun, a 5-inch smoothbore firing a fin-stabilized shell based very much on the Peenemunde arrow shell; this weapon was to be provided with rapid loading from twin rotary magazines, and was to have a water-cooled barrel in order to try and control the wear arising from overheating due to the high rate of fire envisaged.

(*Above*) A more practical weapon was this 155-mm SP gun, also of American origin. In the foreground are the shell and bagged charge.

(*Top right*) A post-war French design was this amalgamation of the British 17-pounder anti-tank gun, the Lorraine chassis, and an incredibly complicated but highly effective muzzle brake by a M Galliot.

(*Right*) The American M44 155-mm SP howitzer; this is the first of its kind to enter British service, and behind it can be seen a 25-pounder 'Sexton' which it replaced.

This gun, known as 'Green Mace' could reach 90 rounds per minute, and another version, using a conveyor-belt feeding ammunition from an 800-round mobile magazine parked alongside the gun, was proposed. But before this weapon could complete its firing trials, a workable and reliable anti-aircraft missile was introduced into service, and overnight the heavy anti-aircraft gun vanished from British service.

The Americans had begun their programme to re-equip the field forces during the war, with an ambitious design of 75-mm gun using a rotary magazine and carrying its own radar and predictor on the mounting. The weapon was also designed around the proximity fuze, in that 75 mm was selected as the smallest-calibre shell which could use this type of fuze and still have a good lethal effect (for proximity fuzes were larger than normal fuzes and occupied some of the interior of the shell normally filled with explosive), and the loading cycle was designed without provision for fuze setting, unnecessary where proximity fuzes were standardized. Begun as the T22 in August 1944, the gun was built by January 1945; trials were carried on for some years, using different barrel lengths and rifling, before it was finally standardized and issued as the M51 in the early 1950s. Since its purpose in life was to accompany the Army in the field as a light and fast-firing weapon, it survived longer than the British designs, not being superseded by missiles until the middle 1960s.

Today the only anti-aircraft guns in use by the

A line-up of American SP guns at a demonstration in 1945. From front to rear, the 240-mm T92 howitzer; the 8-inch M43; the 155-mm gun M40; the 155-mm howitzer M41 and the 105-mm howitzer M37.

major powers are lightweight weapons for field defence against fast-moving, low-level raiders, and these are likely to disappear in the near future with the perfection of lightweight missile systems such as 'Rapier' and 'Blowpipe'.

In 1944 a German laboratory developed a device code-named 'Pfeifenkopf', which was a simple rocket designed to be fired against tanks, but with a rather sophisticated head which carried a scanning device; this signalled back a dim picture of the target to allow an operator to guide the rocket

to the enemy tank. Another version, 'Steinbocke' was under development, using infra-red detectors which allowed the missile to home automatically on the tank without the assistance of an operator. Both these devices were far from being serviceable when the war ended, but what trials had been done showed that there was definite promise in both ideas, given somewhat better instrumentation, and since the end of the war this field had seen far-reaching developments, to the point where the heavy anti-tank gun has vanished from

Another view of the US 155-mm howitzer M44; notice that the driver sits up alongside the gun, giving him an excellent viewpoint.

The German 75-mm light
gun Model 40, the first
recoilless gun to see
extensive war service. The
sliding block breech carries
the venturi tube, and due to
the absence of recoil the
carriage can be exceptionally light.

Lightness carried to the extreme; the French Delahaye 'Jeep' carrying a rocket launcher, a combination of mobility and fire-power better than some of the early tanks.

French AMX 155-mm self-propelled gun. Protection has been abandoned for lightness, but stability is unimpaired, due to the rear spades.

The American 105-mm howitzer M37, a replacement for the M8 'Priest'. This was approved in 1944 but production did not commence until late 1945.

ORD. 20432

STRICTED

the battlefield. The only guns remaining are also war-babies; one of the more important artillery developments of the war–although it didn't look like it at the time–was the recoilless gun, and to-day the only alternative to a missile for anti-tank work is one of these devices.

The recoilless gun relies on discharging the blast from a portion of its propelling charge to the rear, thus balancing the recoil due to the discharge of the projectile forward. First applied in the war by the German airborne troops in Crete, their 75-mm and 105-mm guns were simply lightweight field guns, taking advantage of the fact that with recoil eliminated the gun carriage can be light, due to the absence of firing stresses. For airborne and mountain operations a lightweight weapon was

The breech and venturis of the British 7·2-inch recoilless gun.

highly desirable, but these weapons were short-barrelled and not particularly accurate at short range, and their application as anti-tank guns was not followed up. There was also the considerable disadvantage of the back-blast which revealed the position of the gun with the first shot, and in such a case the gun has to kill the tank with that first shot, otherwise its chance of firing a second shot is problematical. In Britain, however, recoilless guns were developed with long barrels, better accuracy and a tank-killing shell right from the start; unfortunately that start was somewhat delayed so that weapons were not ready for service when the war ended, but numbers were later issued to infantry units for assessment. The Americans also developed recoilless guns on similar lines and managed to get a number into action in the South Pacific before the war ended, again largely for assessment. The general opinion was that provided the gun was accurate enough to give a good assurance of a first-round hit, and provided the projectile was sufficiently lethal to knock out the tank with the first hit, then the recoilless gun was acceptable. The wartime designs were revised and by the early 1950s the recoilless had completely ousted the conventional gun in the anti-tank role. Its advantages were purely infantry-oriented: it was light, simple, and effective at short range; the shell was generally either a hollow charge or the newer (and more lethal) 'Squash-head', a thin-walled shell filled with plastic explosive which plastered itself on to the tank armour, detonated, and either blew a hole in the armour or blew a sizeable scab off the inside face of the plate. Provided the shell could be made to squash, the effect was devastating; thus it was ideally suited to the recoilless gun, since the velocity was too low to be able to use tungsten-cored projectiles.

With all this design endeavour (and money)

being poured into anti-aircraft and anti-tank designs, it is hardly surprising to find that very little work was done on field artillery weapons for some years after the war; indeed it was the rueful complaint of British field gunners that from 1945 to 1958 they only received one item of new equipment—a thermometer for measuring charge temperatures, and even that was more or less a copy of a wartime American design. The Korean War was fought with the same weapons which had fought the Second World War and indeed much of the ammunition was of wartime manufacture. It was not until the end of the 1950s that it became apparent that the guns of the 1930s were no longer suitable and work began on providing replacements.

(*Right*) A 40-mm 'Bristol Bofors' at practice; this was a post-war modification of the original Bofors to power operation.

British 7·2-inch recoilless gun, developed expressly for demolishing the fortifications expected in France in 1944 but never used.

Here the work was coloured by the atomic threat; no longer was it considered wise to allow the gunners to kneel around their gun in the open air, exposed to flash, radiation and fallout. The first British design proposed to replace the 25-pounder was an 88-mm gun developed by a company who had no previous experience in the gun-designing field and who had been given a specifica-tion and allowed to come up with an answer in their own way; the result was a remarkable-looking weapon, but one which was quite prac-tical and which impressed the soldiers. The 'Garrington Gun' (so-called after its designers) used a platform similar to the 25-pounder, but arched the trail over the head of the detachment so that it supported a glass-fibre shield. This was

not only splinter-proof, but radiation-proof as well, and curtains could be lowered from the inner edge of the shield to provide full protection against all the ill-effects of an atomic bomb in the next county.

In another approach, the 25-pounder was also mounted into a fully protected self-propelled mounting derived from the Centurion tank chassis, but the general opinion was that there was too much chassis for too little gun, and the designers took it away and produced another model, this time using the 5·5-inch medium gun, a much more serviceable proposition. But both these weapons, the 88-mm and the 5·5-inch SP, were to become victims of policy; by the time they were in prototype form, there had been NATO agreement about standard calibres, and neither 88 mm nor 5·5 inches fitted into this scheme. Metrication was the order of the day, and the future field guns of NATO nations would be of 105-mm calibre, and the future medium guns of 155-mm. Since neither the 88-mm nor the 5·5-inch could be reworked into these calibres, they were abandoned. In place

of the 5·5-inch, Britain adopted the American 155-mm self-propelled howitzer M44, and in place of the 88-mm gun an Italian 105-mm pack howitzer was adopted to serve until a British 105-mm weapon could be developed.

The eventual British weapon was the 105-mm self-propelled gun 'Abbot', first put on paper in 1958 and introduced into service in 1964. A fully enclosed and turreted weapon, capable of a respectable turn of speed, with the ability to fire at angles from point-blank to 70 degrees elevation, with a maximum range of over 17,000 metres, Abbot has proved to be one of the best and most versatile self-propelled guns ever designed.

The Americans had also developed a number of models of fully enclosed 105-mm self-propelled howitzers, together with larger weapons. 155-mm and 8-inch howitzers and guns became self-propelled in a variety of models, some with turrets and some without. But the target towards which the American designers were working was the development of a gun which could fire an atomic shell.

The American 8-inch howitzer M43, self-propelled. Introduced just as the war was ending, it was to remain in service for several years.

Drill on a British 95-mm recoilless howitzer in 1945. This weapon was developed as possible equipment for airborne artillery but did not survive the war's end. Notice how the detachment are positioned so as to be clear of the blast from the jet nozzles.

The venturis of the British 3·7-inch recoilless gun.

This development had its roots in a wartime idea to produce two weapons, a 240-mm gun and a 280-mm howitzer, which could be easily moved by road. After studying all the current designs of heavy weapons, the designers came up with the idea of mounting the gun into a simple rectangular box which could be transported by carrying each end on a specially designed motor tractor. To place the gun in action the box was lowered from the trucks to rest on a small circular platform, around which it could be revolved to give 360 degrees of traverse; once pointed, the rear end of the trail was lowered down to rest on a ground float, a small amount of traverse across the float being available for fine adjustment. To lighten the load, the German idea of dual recoil was adopted; the gun recoiled in its cradle in the normal way, and the whole mounting recoiled across the plat-

form and trail float as well, giving an exceptionally stable platform. The original idea of a matched pair of weapons was dropped after the war, and work was concentrated on a single weapon, a 280-mm gun. At the same time work began on the design of a nuclear projectile; the calibre of this new weapon appeared to give promise of being able to fit all the mysteries of the nuclear weapon into the restricted volume of the shell.

The project finally reached a successful conclusion; the gun was built and entered service as the M65, and the atomic projectile was perfected and fired. Numbers of guns were sent to Germany, but once there, certain drawbacks made themselves felt. The 47-ton weapon was no easy proposition to pilot around the countryside, and the task of concealing it on the move or when emplaced was almost impossible. Moreover its maximum range of 31,400 yards was derisory when compared with the 'Corporal' missile or the 'Honest John' free-flight rocket, both of which could sport atomic warheads of greater power than the 280-mm shell. And so another cannon fell before the advance of the missiles.

However, all is not yet lost. Recent years have seen design progressing in another direction. The general trend of thought immediately after the war was to develop weapons for another major conflict, and all thought was dominated by the prospects of an all-out nuclear war. But this threat had now receded, while other activities on the earth's surface have emphasized the need to have a highly mobile artillery capable of rapid deployment in any direction at a moment's notice in order to stamp out a minor conflagration before it gets big enough to warrant the use of the more sophisticated and destructive weapons in the armoury.

(*Left*) When it is necessary to shoot at high angles with a gun not originally designed for that task, one begins by digging a pit and heaving the gun into it, but having got it there (*below*), laying and loading become difficult.

(*Right*) As a result the 25-pounder was re-designed with a trail which hinged in the middle to allow an extra 30 degrees of elevation. The joints are, in this picture, wrapped in cloth to keep dirt and soil out.

Undoubtedly the lightest
self-propelled gun was this
75-mm recoilless rifle on a
scooter, a post-war French
aberration which would
have been more at home in
a James Bond film.

Intended for airborne work,
the Alecto SP gun,
carrying a 95-mm howitzer,
was one of the lightest
self-propelled guns ever
built.

American 155-mm M40 SP guns in action in Korea. This also shows the gunners digging slit trenches for local defence of the position.

As a result the last few years have seen much work being done on the development of lightweight field artillery capable of being transported by air, either by being stowed inside a transport aircraft or slung beneath a helicopter. The Americans began the trend by lightening their 105-mm howitzer and redesigning it to become the M102, and the British, in 1972, unveiled their '105-mm Light Gun' which uses a lightweight tubular trail and is shorn of all possible weight-making luxuries. Firing the same ammunition as Abbot, the light gun can be helicopter-lifted anywhere in

Loading a static-mounted 3·7-inch anti-aircraft gun; each complete round weighs fifty pounds and twenty could be fired in one minute.

British officer students firing anti-tank practice in 1954. Bren light machine-guns are clamped to the barrels of the 25-pounder and connected to the gun firing lever, allowing short-range practice using ·303 tracer bullets.

Rear view of the 175-mm M107 showing the massive recoil spade and the power rammer assembly folded up to the left of the breech.

A British 25-pounder recoils; an unusual picture, since the 'gunner' at the right is actually a Grenadier Guardsman.

(*Bottom right*) A 175-mm M107 fires, raising the dust. The barrel is wrapped to provide camouflage and also to minimize the chance of detection by infra-red devices.

the world, can be manœuvred in difficult country by manpower and, as the Americans showed with their model in Vietnam, can even be floated on platforms in swamps to fire from positions previously considered impossible.

What of naval gunnery? This, too, has seen a resurgence. The war, with its accent upon air power and the inevitable postwar atomic thinking, led numerous observers to the conclusion that the day of the naval gun was over. Much postwar development would appear to reinforce

this opinion, with the adoption of missile-carrying submarines and deck-carried missiles for surface craft. Nevertheless, there is still room for the gun; a point which was well brought out when the Vietnam War brought a demand for ship-to-shore bombardment and the US Navy had a hectic time de-mothballing gun-armed vessels to go and do what missile-armed warships couldn't–shell a hostile shoreline.

By and large the current trend in naval ordnance is toward dual-purpose weapons which

An American Army 155-mm M109 bustles across the North German plain on an exercise. Notice the prominent fume extractor and muzzle brake.

The writing on the wall; a 'Corporal' missile, warhead gleaming in the sunshine, is prepared for firing.

can function as ship-to-ship or ship-to-air weapons; this demands a rapid-firing capability, together with turret mountings which protect the gunners from the usual post-atomic problems of fall-out and radiation. In British service much of the development of fast-firing weapons came as a spin-off from the rapid-loading designs for anti-aircraft guns, so that the long development of the 'Ratefixer' programme was far from being entirely wasted. A 3-inch rapid-fire gun was first produced, to be followed soon after by a 6-inch, and similar weapons have been adopted by the US Navy. The mechanical glamour of a three-gun 15-inch turret may be missing, but the firepower of a modern warship, even discounting whatever may be available in the missile cupboard, is as formidable as most of its forebears.

Much the same can be said of land artillery: the advent of improved explosives, perfected designs of shell, rapid loading devices, electronic computers to calculate firing data, more electronic devices to detect targets, assess them and feed the information into an automatic data processing unit, all adds up to a response time which is incredibly short and a firepower capability and accuracy which is far and away superior to the artillery of yesteryear. We may not have the enormous eye-catching railway guns, the ponderous siege artillery, the mechanically ingenious coast mountings or the sleek and threatening anti-aircraft and anti-tank guns of 30 years ago, but when it comes to the basic mission, today's infantrymen and cavalrymen, riding to war in their armoured chariots, are assured of artillery support as effective as anything their forefathers ever had — or hoped for. Prinz Kraft's basic dictum still applies:

(1) First it must *hit*; second, *hit*; and third *hit* . . .
(2) It must be in the right place at the right time . . .

Index

Page numbers in *italic* denote illustrations

'Abbot', British 105-mm SP Gun 228
Alanbrook, FM Lord 153
American:
  artillery production during First World War
    157
  boat gun *67*
  Civil War 61, *67*
  early guns 61
  ordnance:
    Armstrong Whitworth *56*
    boat gun *67*
    Columbiad *72*
    Dahlgren 11-inch on *Monitor 118*
    Parrott 20-pounder *56*
    Rodman 15-inch *59, 87*
    Zalinski Dynamite Gun *87, 88*
    75-mm howitzer M1 *190, 196*
    75-mm experimental divisional gun *165*
    3-inch seacoast M1903 *93*
    3-inch AA M1923E *156*
    90-mm anti-tank T8 *182*
    105-mm howitzer M1 *164, 175*
    105-mm howitzer on half-track *177*
    105-mm self-propelled howitzer M37
      222, 225
    4·7-inch AA gun M1923, self-propelled
      146
    155-mm gun M1 'Long Tom' *192*
    155-mm self-propelled gun M40 *220,
      233*
    155-mm howitzer self-propelled M41
      222
    155-mm howitzer self-propelled M44
      221, 222
    155-mm howitzer self-propelled M109
      236
    175-mm gun self-propelled M107 *234,
      235*
    7-inch siege howitzer M1890 *106*
    8-inch gun M1 *188–9*
    8-inch howitzer self-propelled M43 *222,
      228*
    240-mm howitzer M1918 in coast role
      154
    240-mm howitzer M1 *197*
    240-mm howitzer self-propelled T92
      211, 222
    280-mm gun M65 *229*
    13-inch seacoast mortar *64*
    14-inch seacoast gun M1907 *91*
    16-inch seacoast gun *161*
    36-inch 'Little David' howitzer 209, *212,
      213*
Amusette 40
Anti-aircraft guns:
  first designs 112
  early problems 112
  First World War types 139
  performance during First World War 141
  post-1945 development 218
  Second World War types 189
Anti-submarine weapons 147
Anti-tank ammunition 182
Anti-tank artillery 180
Anti-tank missiles, early types 222
Archimedes 10
Armour-piercing shell 85, 120, 146
Arms & Explosives 125
Armstrong, Sir William 58
Armstrong guns:
  basic features 59
  defects 63
  tested in battle 62
  40-pounder gun *57*
  110-pounder gun *57*
  100-ton gun 89
  7-inch Naval gun *66*
Armstrong balanced pillar mounting *116*

Armstrong hydro-pneumatic disappearing
  mounting *96*
Armstrong protected barbette system *95*
Armstrong-Whitworth breech-loading gun *56*
Armstrong and Whitworth gun trials 63
Arrows as artillery projectiles 15
Artillery, definitions 11
Atmospheric conditions 29
Atomic cannon, US 280-mm gun M65 230
Austro-Hungarian ordnance:
  coast defence gun *ca* 1885 *117*
  30·5-cm howitzer *135*
Austro-Prussian War 61
Autocannon, French 75-mm AA gun 139
Auto-frettage system 107
Auxiliary-propelled guns *181, 182*

Bacon, Roger 11, 12
Balanced pillar mounting, Armstrong's *116*
Ballistic diagram *ca* 1620 *29*
Ballistic measuring machines 48, 74, *82*
Ballistic pendulum 46
Balloon guns 112
Bar shot *70*
Bashforth, Rev 74
Battalion gun 33, *34*
Beaulieu, Colonel Treuille de 57, 60, 69
Beejapore, Great Gun of 23
Berger's cast-steel gun *89*
Bessemer's stabilized gun mounting *75*
Birch, General Sir Noel 163
Birch 18-pounder Mark 5, self-propelled gun
  162, 163
Blenheim, Battle of 33
Bofors guns — see under Swedish Ordnance
Bolimov, Battle of 154
Bomb ship, section *37*
Bonneville, M de 40
Boring machine for cannon *29*
Boulenge, Capt-Commandant P le 74
Bourne, early writer on artillery 28
Boxer, Colonel 84
Boxer's improved shrapnel shell *49*
Brackenbury, Gen Sir Henry 10, 100
Brandt, Edgar 184
Breech-loading guns:
  Armstrong 59
  earliest systems 20
  early naval designs *42*
  German, of 16th Century *27*
  improved systems 60
Bristol Bofors 40-mm light anti-aircraft gun
  227
British Artillery policy 1922 162
British ordnance:
  Armstrong 40-pounder *57*
  Armstrong 110-pounder *57*
  Armstrong 7-inch Naval *66*
  Armstrong protected barbette *95*
  Armstrong balanced pillar *116*
  Elswick Ordnance Co, 6-inch broadside *73*
  Elswick Ordnance Co, 6-inch Mark 9 *90*
  Elswick Ordnance Co, 6-inch disappearing
    112
  Lancaster oval-bore *54*
  Mallet's 36-inch mortar *60*
  Maxim one-pounder pom-pom *140*
  Mons Meg *18*
  Smith Gun *202*
  Vickers 70-mm infantry gun *162*
  Vickers 105-mm field gun *167*
  Vickers 105-mm field howitzer *172*
  Vickers 8-inch howitzer *124*
  2-pounder anti-tank gun *173*
  twin 2-pounder Naval AA gun *203*
  light 6-pounder for Canadian service *35*
  long 12-pounder, Naval *65*
  12-pounder moveable armament *144*

13-pounder Royal Horse Artillery gun *160*
17-pounder anti-tank gun *205*
17-pounder 'Straussler' auxiliary propelled
  181
18-pounder field gun *128, 142, 143, 145,
  152*
18-pounder Mk 5 self-propelled gun *162,
  163*
25-pounder field gun *172, 183, 191, 193,
  230, 231, 235*
25-pounder 'Short' gun *208*
32-pounder smoothbore garrison gun *43*
32-pounder smoothbore Naval gun *65*
32-pounder carronade *72*
32-pounder anti-tank *180*
3-inch 20-cwt anti-aircraft *144, 145*
3·3-inch mock-up *172*
3·6-inch anti-aircraft *167*
3·7-inch anti-aircraft *191, 233*
3·7-inch mountain howitzer *156, 159*
3·7-inch recoilless *229*
4-inch jointed gun *84*
4·5-inch howitzer *150, 151*
4·7-inch coast defence gun *111*
4·7-inch heavy field gun *125*
5·5-inch medium gun *207*
6-inch coast defence *170*
6-inch 80-pounder Naval *130*
6-inch high velocity experimental field *163*
6-inch 26-cwt howitzer *110*
6-inch 30-cwt howitzer *119*
6-inch Naval on centre pivot mounting *120*
7·2-inch howitzer *187*
7·2-inch recoilless *225, 227*
8-inch howitzer *124, 137*
9-inch rifled muzzle loading, high angle *61*
9·2-inch coast defence *95, 111, 115*
9·2-inch howitzer *199*
9·2-inch Naval, Vavasseur mounting *119*
9·2-inch muzzle-loading *125*
10-inch on Easton & Anderson mounting
  97
11-inch 25-ton rifled muzzle loader *73*
12-inch railway gun *126*
12-inch railway howitzer *170*
12-inch road howitzer *133*
13-inch muzzle-loading mortar *51*
14-inch railway gun *127*
15-inch howitzer *134*
20-mm Triple Polsten AA gun *210*
40-mm Bristol Bofors AA *227*
75/120-mm interchangeable-barrel gun
  155
95-mm recoilless gun *229*
95-mm self-propelled howitzer 'Alecto' *232*
105-mm light gun *233*
Broadside mountings 116
Broadwell ring obturation system *59*, 60
Bronze as gun material 27
'Bruno', German 24-cm railway gun *198*
'Bussen' early expression for gun 13

Campbell, Sir Frederick 92
Cannon, table of types and names 28
Capped projectiles 120
Captain, HMS, turret ship 75
Carriages:
  early development 41
  modern designs 108
Carronade 71, *72*
Cartridges, first mentioned 29
Case shot *85*
Casting of cannon, in England 27
Catapult, Chevalier Folard's *39*
Cavalli, Major 57
Central battery system of ship armament 73
Centre pivot mounting *120*
Cerisolles, Battle of 10

Chain shot 70
Charge ignition, methods 87
Christie self-propelled mounting 146
Chronograph, development of 78
Chronoscope, development of 78, 82
Churchill, Sir Winston 128
Coast defence, activity in 19th century 69
Coast defence guns:
  American 13-inch mortar 64
  American 14-inch 92
  American 3-inch rapid fire 93
  Austro-Hungarian 117
  British 9·2-inch 111, 115
  British 6-inch 170
  British 4·7-inch 111
  French 24-cm 63
  French turret 194
  high angle 9-inch rifled muzzle loader 61
  Indian 7·5-inch 106
  Jersey, Castle Coronet 85
  Portuguese 63
Cocoa Powder 80
Coles, Captain Cowper, RN 74
Columbiad gun 71, 72
Common shell 85, 86, 101
Compressor system of controlling recoil 73
Conders' multiple chamber gun 204, 209
Constantinople, Siege of 23
Cordite, development of 83
Corned powder 26
Corporal missile 236
Coventry Ordnance Works 128
Crécy, Battle of 15
Crimean War 52, 54
Croly, Lieut 57
Culverin 32

Dardanelles 23
De Bange system of obturation 90, 142
De Dion autocannon 143
Deport mountain gun 104
Dhool Danee, great gun of Agra 23
Differential recoil 104, 108, 159
Disappearing carriages:
  American experimental 91
  Armstrong hydro-pneumatic 96
  Buffington-Crozier 91, 92, 93
  Easton & Anderson 97
  Elswick Ordnance Company 112
  Moncrieff 91, 93, 94
  Redlichkeit 40
Discarding Sabot projectiles 184
Dismounting of 'tween-deck guns 121
Diver raising sunken cannon; old engraving
  15
Douglas, Sir Charles 88
Duhem, Prof Pierre 12
Dulle Greite, Duke of Burgundy's cannon 22
Dutch 80-mm field gun 96
Duties of a Gunner 29
Dynamite cannon 85, 87, 88

Easton & Anderson's disappearing carriage
  112
Eberhardt, Dr 135
Electric firing of guns 89
Electric guns 203
Elswick Ordnance Company:
  broadside mounting 73
  cup obturation system 90
  6-inch Mark 9 gun of 1901 90
  disappearing carriage 112
  naval turret 114
Engines of war 30, 31
English, Captain, RE 120
Engstrom's gun carriage 58
Erhardt 15-pounder gun 100
Examination of bore by reflected sunlight 12

Falconet, Swiss cannon 31
Firing locks, Naval 89
First World War 124 et seq
'Fish Gun' of the King of Oudh 47
Flash spotting 177
Flintlock cannon ignition 88, 89
Florence, first authentic record of artillery at
  14
Folard, Chevalier 34, 39
Fortress cannon, 14th century 14
Franco-Prussian War 62
Frederick the Great 33, 34
French fortress turret 92
French ordnance:
  12-pounder field artillery, ca 1790 36
  12-pounder horse artillery, ca 1790 35
  17-pounder on Lorraine chassis 221
  37-mm Trench cannon 153
  47-mm Puteaux anti-tank gun 210
  75-mm anti-aircraft gun 157
  75-mm De Dion Autocannon 143
  75-mm Deport mountain gun Mle 1910
    104
  75-mm Field Gun M1897 95, 157, 193,
    209
  75-mm recoilless gun on scooter 232
  75-mm St Chamond mountain gun 126
  75-mm Schneider mountain gun 196
  105-mm gun M1913 176
  120-mm siege gun Mle 1878 109
  155-mm gun Mle GPF 185
  155-mm Schneider howitzer M1917 169
  155-mm self-propelled gun on AMX
    chassis 224
  155-mm siege gun M1877 103
  24-cm coast gun Mle 1876 63
  30·5-cm railway gun 138
  34-cm coast defence turret 194-5
  34-cm railway gun 129
  40-cm railway gun 137
  Richelieu turret 131
Friction tube ignition 88
Froissart 14
Fuzes:
  early development 83
  sections of typical models 132

Galloper guns 37, 40
'Gamma' howitzer 127
Garrington gun 227
Garrison artillery at drill 49, 50-1, 79-82
Gas warfare 154
Gatling gun 113
Gentoo Code 10
Gerlich 181
German ordnance:
  16th century breech-loading gun 27
  16th century mortar 27
  Krupp breech-loading steel gun 58
  Krupp 65-mm anti-aircraft gun of 1910 107
  Krupp 75-mm mountain gun of 1909 159
  Krupp 16-inch 71-ton coast gun of 1879
    90
  75-mm infantry gun 18 165
  75-mm LG 40 recoilless gun 223
  77-mm field gun M1896 160
  77-mm field gun M1916 138
  80-mm PAW 8H63 anti-tank gun 226
  87-mm naval gun ca 1886 147
  10-cm experimental anti-tank gun 214
  105-mm anti-aircraft gun 35, 168
  105-mm field howitzer 18 211
  105-mm field howitzer 18/40 205
  128-mm anti-tank gun K44 180, 181
  15-cm heavy field gun 18 202
  15-cm heavy field howitzer 18 186
  15-cm howitzer 68
  17-cm gun 186

  21-cm gun K38 201
  21-cm long range gun K12 168, 198
  24-cm railway gun 'Bruno' 198, 199
  28-cm railway gun K5 199
  38-cm coast defence gun 141
  38-cm railway gun 'Max' 136
  60-cm self-propelled gun 'Karl' 207
  80-cm railway gun 'Gustav' 200, 207
German re-armament in 1930s 167
Ghent, Memorialbuch der Stadt 13
Gibraltar, development of shrapnel shell at 51
Grafton's Chronicles 13
Grapeshot 47
Grapnel shot 70
Green Mace anti-aircraft gun 220
Gribeauval 50
Gruson 92, 94
Gun construction:
  auto-frettaged 107
  built-up 105
  longitudinal bars 14, 16
  single tube 106
  stages in 19-21, 28
  wire-wound 105
Guncotton, invention of 83
Gunner's Quadrant 27, 41
Gunpowder:
  corning of 26
  cost of 19
  defects of 26-7
  later improvements 80
  strength and composition 16
  serpentine 26
  table of service grades 82
'Gustav', German 80-cm railway gun 207
Gustavus Adolphus 30

Half-track mounting of US 105-mm howitzer
  177
Halle, Sebastian 43
Headlam, Major-General Sir John 150
Helena, USS 215
High angle coast defence guns 61
High explosive shells developed 85
Hime, Lt-Col H W L 11
Hindersen, General von 62
Hollow charge anti-tank shell 189
Horse artillery:
  in British service 50
  systems of 50
Horse teams 50
Hypervelocity guns 201

Interchangeable barrel gun 155
Intermediate anti-aircraft guns 198
Interrupted screw breech mechanism 108,
  141, 146
Ironclad warships 64
Iron ship armour construction 62
Italian ordnance:
  14th-century gun 30
  15th-century culverin 32
  15th-century siege mortar 117
  75-mm field gun 204

Japanese ordnance:
  75-mm Mountain gun Model 94 209
  24-cm siege howitzer 130
Jointed gun 84
Jutland, Battle of 146

Kaiser Wilhelm Geschutz 135
Kaiser Wilhelm Institute, Berlin 125, 154
'Karl', German 60-cm self-propelled howitzer
  207
Kohler's depression carriage 42
Korean War 226
Kraft zu Hohenloh-Ingelfingen, Prince 62

Krupp:
    anti-aircraft gun 1910 *107*
    breech closing systems *58*, 60
    develops steel ordnance 60
    early BL guns in Boer War *58*
    'Grasshopper' weapons carrier 219
    howitzers at Liège 126
    mountain gun with differential recoil *159*
    offers gun on trial to Britain 91
    16-inch gun of 1879 *90*

La Gloire, French ironclad 64
Lancaster system of rifled ordnance *54*, 58
'Lange Emil', German 38-cm gun 134
Leather guns 31
Le Cateau, Royal Artillery at 150
Leipsig, Battle of 31
Liège Forts, German attack on 126
'Little David', US 36-inch howitzer 209, *212–13*
Long 12-pounder naval gun *65*
Longhand, anti-aircraft gun development programme 219
Long range gun, German 21-cm K12 168
Lucar, Cyprian 28–9
Lyman and Haskell multiple chambered gun 204

McIntosh's recoil carriage patent *75*
Makarov, Admiral 120
Malplaquet, Battle of 33
Marignan, Battle of 10
Marlborough, Duke of 33
Marshall, Commander 74
'Max', German 38-cm railway gun *136*
Maxim one-pounder pom-pom *140*
Mechanical traction, early attempts at 115
Mefford, and pneumatic gun *85*, *87*
Mercenary troops, employment of 23
Millimete, Walter de 15
Millimete gun *13*
Misfire of 100-ton gun at Gibraltar 90
Missouri USS, turret accident in 116
Mohammed, artillery of *19*, 23
Monarch, HMS, turret ship 74
Moncrieff, Captain *91*, *93*, *94*
Monitor, USS, section *118*
Mons Meg *18*
Mortars:
    as siege weapons 54
    battery of, in action, 18th century *23*
    German, 16th century *27*
    Mallet's 36-inch 60
    naval *36*, *37*
Moscow, Great Gun of 23
Mould for cannon balls *29*
Mountain artillery 52
Muk el Maiden, great gun 23
Muzzle brakes:
    Col de Beaulieu's pattern *69*
    de Place's pattern *139*
Muzzle pivoting carriage, Shaw's *75*

Napoleon's *Artillery* 14
NATO standardization of calibres 228
Naval fire control 118
Naval ordnance:
    broadside cannon *38*
    early breechloader *42*
    long 12-pounder gun *65*
    mortar *36*
    truck carriage *38*
    twin 2-pounder anti-aircraft *203*
    32-pounder gun *65*
    32-pounder Carronade *72*
    6-inch 80-pounder of 1882 *130*
    6-inch central pivot *120*
    9·2-inch on Vavasseur mounting *119*

11-inch 25-ton on *Temeraire* mounting slide *73*
87-mm German *ca* 1886 *147*
Richelieu turret *131*
Naval turrets *114*, 116, *131*
Navez, Major 78
Neuve Chapelle, Battle of 151
Noble, Sir Andrew 78, *82*, 138
Noble, Lieut W H 70
Nordenfelt 57-mm gun in Russian service *104*

Obturation systems 90
Obuchov, Russian 24-pounder gun *54*
Odruik, Siege of 16
Opus Tertium, manuscript of Roger Bacon 12
Oudenarde:
    Battle of, 1708 33
    Siege of, 1382 19

Pacific Ocean, defensive works in 165
Paixhans, General 72
Palliser shells 71, *120*
Palliser system of re-lining smoothbore guns *64*, 69
Paris Gun 135
Parrott, Robert P *56*, 61
Peenemunde Arrow shell 199
Peerless lorry mounting for anti-aircraft gun *144*
Pershing, General 161
Peterara, early breech-loading cannon *17*, 20
Piobert 54
Pom-pom, Maxim one-pounder 139
Port Arthur, Siege of 126
Portuguese ordnance:
    breech-loading field gun *60*
    coast gun on recoil slide *63*
    15-cm Skoda field gun *110*
Pot-de-fer 16
Poudre, B 83
Prism gunpowder 81
Proximity fuze 214
Prussian army and steel guns 60
Prusso-Danish War 61
Puff, Karl 181
'Punching' system of armour attack 70
Pyro powder 83

Quesnoy, Siege of 14
Quill tubes 88

'Racking' system of armour attack 70
Railway guns:
    general 130
    British: 9·2-inch Mk 9 *125*
        12-inch Mk 9 *126*
        12-inch Mk 3 *170*
        14-inch Mk 1 *127*
    French: 30·5-cm *138–9*
        34-cm *129*
        40-cm *137*
    German: 24-cm 'Bruno' *198–9*
        28-cm K5 199
        38-cm 'Max' *136*
        80-cm 'Gustav' *200*
Ram, as a naval weapon 73
Ramsay, Capt Norman 52
Rangefinder, operation on *121*
'Ratefixer', anti-aircraft gun development programme 219
Rausenberger, Professor 128, 135
Recoil:
    problems and systems of control 94
    hydro-pneumatic and hydrospring systems 107
Recoilless guns *223*, *225*, *227*, *229*, *232*
Redhot shot 41
Redlichkeit's disappearing carriage *40*

'Rendable' shell 121
Ribaudequin *16*, 19
Richelieu turret *131*
Richmond, Duke of 50
Rifled muzzle loading adopted 64
Rifled ordnance, advantages 60
Rifling machine 59
Rihoult, castle of, inventory of weapons 15
Ring shell *62*
Robins, Benjamin 46–7
Rocket assisted shell 199, *201*
Rocket launcher, German 30-cm *206*
Rodman, Capt T J 61, 78
Rodman gun, 15-inch *59*, *87*
Rolf Krake, first turret ship 74
Rouen, marine arsenal at 14
Russian ordnance:
    'Obuchov' gun *54*
    'Schuwalon' gun *41*
    'Tsar Puschka' *18*
    45-mm anti-tank M1942 *176*
    57-mm Nordenfelt *104*, *109*
    6-inch M1878 *102*
Russo-Japanese War 100

St Chamond mountain gun *126*, *127*
St Omer 14
San Sebastian 52
Sawyer rifled shell *53*
Saxe, Marshal 40
Schneider-Canet Company 130
Schonbein, Prof 83
Schultz powder 83
Schuwalon gun *41*
Schwartz, Berthold 13
Sedgemoor, Battle of 32
Segment shell 84
Self-propelled guns:
    British policy 163
    developments during Second World War 208
    early designs 143–6
    Alecto 95-mm *232*
    Birch 18-pounder *162*, *163*
    Bumble Bee *218*
    Christie *146*
    Grizzly Bear *219*
    Holt *153*
    Hornet *215*
    'Karl' *207*
    Sexton 208
    3·7-inch anti-aircraft *197*
    240-mm howitzer *211*
Semi-armour-piercing shells 121
Serpentine powder 26
Shaw's muzzle-pivoting carriage *75*
Shells:
    common 101
    for rifled ordnance *53*
    introduced 42
    Palliser 71, *120*
    Shrapnel 51, 84, *86*, 101
    smoothbore types *47*
    Star *101*
Shimonoseki, Battle of 105
Ship carriages, early designs 43
Shrapnel, Lieut Henry 50–51
Siege warfare 33
Sighting systems 110
Singapore, defences of 166
Sliding block breech mechanism *107*, *152*
Sliding railway mounting 133
Smith gun *202*
Smith-Dorrien, General Sir Horace 150
Snow, General Wm J 156
Soft recoil 108
Sound cannon 203
Sound ranging 177

South African War 100
Spanish ordnance:
  cannon of 1628 *43*
  15-cm howitzer M1887 *105*
Special Committee on Iron 67
Spherical case shot *47*, 51
Sponson system of ship armament 73
Spring spade control of recoil 95
Squash-head shell 226
Squeezebore projectiles *183*
Stabilized gun mounting, Bessemer's patent
  75
Stone shot vis-à-vis iron 19
Straussler auxiliary-propelled gun *181*
Summerall, Col Charles P 161
Surinam, shrapnel shell used at 51
Swedish ordnance:
  leather guns *31*
  Bofors: 37-mm anti-tank *175*
    40-mm anti-aircraft *190*, 192
    80-mm anti-aircraft *166*
    90-mm field *169*
    105-mm field *164*
    12-cm howitzer *105*
Swiss Falconet of 1672 *31*
Systems of horse artillery 50

Tactics, in Middle Ages 26
Taper-bore guns 181
Tartaglia, Nicolo 28, 30
Temeraire, HMS:
  and disappearing carriage 93
  11-inch 25-ton gun and slide mounting *73*
Three-barrel cannon *26*
Thunderer, HMS:
  accident with RML gun on board 92
  turret for breech-loading guns *114*
Tippoo Sahib, mortar of *17*
Traclat, experimental artillery tractor *185*
Trench cannon, 37-mm *153*
Triple-shot cannon *11*
Truck carriage *38*, 43
'Tsar Puschka', great gun of Moscow *18*, 23
Tungsten, use in piercing projectiles 183
Turkish cannon 23
Turrets, naval:
  Elswick Ordnance Company *114*
  HMS *Thunderer 114*
  Richelieu 131

Universal shell 100

Vavasseur mounting *119*
Vesuvius, USS, armed with dynamite cannon
  86
Vickers:
  8-inch howitzer *124*
  70-mm infantry gun *162*
  105-mm field gun *167*
  105-mm field howitzer *172*
Victory, HMS, section *68*

Wahrendorff, Baron 57
Warrior, HMS 64
Washington Conference 165
Wheatstone, Prof 78
Whirlwind cannon 202
Whitworth, Joseph 63
Whitworth shell 53
William of Orange 32
Wirewound system of gun construction 105
Wolseley, General Sir Garnet 90
'Woolwich system' of rifled ordnance 69

Zalinski Dynamite Gun 85, *87*, *88*
Zippermeyer's Vortex cannon 202

Acknowledgments

The publishers are grateful for the
use of illustrations from the author's collection
and from the following sources:

Alexander Turnbull Library
Brian J. Hebditch
Imperial War Museum
Inspector-General of Armaments
Royal Artillery Institution
School of Artillery
Services Americains D'Information
G. Z. Trebinski
US Signal Corps, National Archives
Washington States Parks
and Recreation Commission